MANUFACTURING TECHNOLOGY RESEARCH

UNITED STATES MANUFACTURING

PERSPECTIVES AND PROSPECTS

Manufacturing Technology Research

Additional books in this series can be found on Nova's website under the Series tab.

Additional E-books in this series can be found on Nova's website under the E-book tab.

Economic Issues, Problems and Perspectives

Additional books in this series can be found on Nova's website under the Series tab.

Additional E-books in this series can be found on Nova's website under the E-book tab.

MANUFACTURING TECHNOLOGY RESEARCH

UNITED STATES MANUFACTURING

PERSPECTIVES AND PROSPECTS

VANCE MURPHY
AND
JILLIAN E. WALKER
EDITORS

Nova Science Publishers, Inc.
New York

For permission to use material from this book please contact us:
Telephone 631-231-7269; Fax 631-231-8175
Web Site: http://www.novapublishers.com

Library of Congress Cataloging-in-Publication Data

ISBN: 978-1-62081-522-9

Published by Nova Science Publishers, Inc. † New York

CONTENTS

PREFACE

The health of the U.S. manufacturing sector has long been of great concern to Congress. The decline in manufacturing employment since the start of the 21st century has stimulated particular congressional interest. Members have introduced hundreds of bills intended to support domestic manufacturing activity in various ways. The proponents of such measures frequently contend that the United States is by various measure falling behind other countries in manufacturing, and they argue that this relative decline can be mitigated or reversed by government policy. This book informs the debate over the health of U.S. manufacturing through a series of charts and tables that depict the position of the United States relative to other countries according to various metrics. Understanding which trends in manufacturing reflect factors that may be unique to the United States and which are related to broader changes in technology or consumer preferences may be helpful in formulating policies intended to aid firms or workers engaged in manufacturing activity.

Chapter 1 - The health of the U.S. manufacturing sector has long been of great concern to Congress. The decline in manufacturing employment since the start of the 21st century has stimulated particular congressional interest. Members have introduced hundreds of bills intended to support domestic manufacturing activity in various ways. The proponents of such measures frequently contend that the United States is by various measures falling behind other countries in manufacturing, and they argue that this relative decline can be mitigated or reversed by government policy.

This report is designed to inform the debate over the health of U.S. manufacturing through a series of charts and tables that depict the position of the United States relative to other countries according to various metrics. Understanding which trends in manufacturing reflect factors that may be unique to the United States and which are related to broader changes in technology or consumer preferences may be helpful in formulating policies intended to aid firms or workers engaged in manufacturing activity. This report does not describe or discuss specific policy options.

Chapter 2 - The health of the U.S. manufacturing sector has been a long-standing concern of Congress. Although Congress has established a wide variety of tax preferences, direct subsidies, import restraints, and other federal programs with the goal of retaining or recapturing manufacturing jobs, only a small proportion of U.S. workers is now employed in factories. Meanwhile, U.S. factories have stepped up production of goods that require high technological sophistication but relatively little direct labor. Labor productivity in manufacturing, as measured by government data, has grown rapidly, suggesting that the manufacturing sector as a whole remains healthy.

Recent data, however, challenge the belief that the manufacturing sector, taken as a whole, will continue to flourish. Unlike previous expansions, the two most recent cyclical upturns in the U.S. economy have not generated jobs in manufacturing. Moreover, statistics suggest that domestic value represents a diminishing share of the value of U.S. factory output. One interpretation of these data is that manufacturing is "hollowing out" as companies undertake a larger proportion of their high-value work abroad. These developments raise the question of whether the United States will continue to generate highly skilled, high-wage jobs related to advanced manufacturing.

Chapter 3 - Exporting provides opportunities to expand the sales of U.S. manufacturing industries, resulting in greater revenues for U.S. companies and higher employment for U.S. workers. However, investors and workers are concerned about risk as well as expected return, and risk depends on the volatility of economic outcomes. In the case of exports, it is worth asking whether the added risks and unpredictability associated with exporting outweigh the increase in expected revenues.

In theory, exporting could increase the volatility of the total shipments of U.S. manufacturers by increasing cost competition and by exposing the manufacturers to other types of uncertainty, including exchange rate volatility and fluctuations in shipping costs. On the other hand, exporting could reduce the volatility of their total shipments by diversifying shocks to product demand.

In this paper, we investigate the recent link between exporting and the economic stability of the U.S. manufacturing sector. We compare the volatility of the domestic shipments of U.S. companies and industries to the volatility of their total, worldwide shipments.

Chapter 4 - The United States has long thrived as a result of its ability to manufacture goods and sell them to global markets. Manufacturing activity has supported our economic growth, leading the Nation's exports and employing millions of Americans. The manufacturing sector has also driven knowledge production and innovation in the United States, by supporting two-thirds of private sector research and development and by employing scientists, engineers, and technicians to invent new products and introduce innovations in existing industries.

The Nation's historic leadership in manufacturing, however, is at risk. Manufacturing as a share of national income has declined, as has manufacturing employment, and our leadership in producing and exporting manufactured goods is in question. The loss of U.S. leadership in manufacturing, moreover, is not limited to low-wage jobs in low-tech industries, nor is it limited to our status relative to low-wage nations. The United States is lagging behind in innovation in its manufacturing sector relative to high-wage nations such as Germany and Japan, and has relinquished leadership in high-tech industries that employ highly-skilled workers. Our trade balance in advanced technology manufactured products—long a relative strength of the United States—shifted from surplus to deficit starting in 2001,[1] and a trade deficit of $17 billion in 2003 further widened to $81 billion by 2010.[2] In addition, the United States has been steadily losing the research and development activity linked to manufacturing—and associated high-skilled jobs—to other nations, as well as our ability to compete in the manufacturing of products that were invented and innovated here—from laptop computers to flat panel displays and lithium ion batteries.

Chapter 5 - Declining U.S. manufacturing has been an issue of continuing concern for policymakers; this was reflected in the Obama Administration's (Administration) 2010 announcement of the National Export Initiative. The Administration has also shown interest in improving the efficiency of the federal support of trade operations. In 2004, the Office of Manufacturing and Services (MAS) was established within the Department of Commerce's (Commerce) International Trade Administration (ITA) to enhance the global competitiveness of U.S. industry.

GAO was asked to examine (1) MAS's goals and activities and how they compare with those of other government entities; (2) how MAS prioritizes its activities and targets its resources; and (3) the extent to which MAS tracks and reports its efforts. GAO reviewed agency documents and interviewed officials from MAS, other parts of ITA and Commerce, and other agencies.

In: United States Manufacturing
Editors: V. Murphy and J. E. Walker
ISBN: 978-1-62081-522-9

Chapter 1

UNITED STATES MANUFACTURING IN INTERNATIONAL PERSPECTIVE*

Marc Levinson

SUMMARY

The health of the U.S. manufacturing sector has long been of great concern to Congress. The decline in manufacturing employment since the start of the 21st century has stimulated particular congressional interest. Members have introduced hundreds of bills intended to support domestic manufacturing activity in various ways. The proponents of such measures frequently contend that the United States is by various measures falling behind other countries in manufacturing, and they argue that this relative decline can be mitigated or reversed by government policy.

This report is designed to inform the debate over the health of U.S. manufacturing through a series of charts and tables that depict the position of the United States relative to other countries according to various metrics. Understanding which trends in manufacturing reflect factors that may be unique to the United States and which are related to broader changes in technology or consumer preferences may be helpful in formulating policies intended to aid firms or workers engaged in manufacturing activity. This report does not describe or discuss specific policy options.

The main findings are:

* This is an edited, reformatted and augmented version of Congressional Research Service, Publication No. R42135, dated January 5, 2012.

- The United States remained the largest manufacturing country in 2010, although its share of global manufacturing activity has declined in recent years.
- Manufacturing output has grown more rapidly in the United States over the past decade than in most European countries and Japan, although it has lagged China, Korea, and other countries in Asia.
- Employment in manufacturing has fallen in most major manufacturing countries over the past two decades. The United States saw a disproportionately large drop between 2000 and 2010, but its decline in manufacturing employment since 1990 is in line with the changes in several European countries and Japan.
- U.S. manufacturers spend far more on research and development (R&D) than those in any other country, but manufacturers' R&D spending is rising more rapidly in China, Korea, Mexico, and Taiwan.
- A large share of manufacturing R&D in the United States takes place in high-technology sectors, particularly pharmaceutical and electronic instrument manufacturing, whereas in other countries a far greater proportion of manufacturers' R&D outlays occur in medium-technology sectors such as motor vehicle and machinery manufacturing.

INTRODUCTION

The health of the U.S. manufacturing sector has long been of great concern to Congress. The large decline in manufacturing employment since the start of the twenty-first century has stimulated particular congressional interest. Members have introduced hundreds of bills intended to support domestic manufacturing activity in various ways. The proponents of such measures frequently contend that the United States is in some way falling behind other countries in manufacturing, and argue that this relative decline can be mitigated by government policy.

Examining U.S. manufacturing in isolation sheds little light on the causes of changes in the structure of the manufacturing sector. While some of those changes may be a result of factors specific to the United States, others may be attributable to technological advances, changed consumer preferences, or macroeconomic forces such as exchange-rate shifts. This report is designed to inform the debate over the health of U.S. manufacturing by examining recent changes in the manufacturing sector in comparative perspective. It does not describe or discuss specific policy options.

The charts and tables on the pages that follow depict the position of the United States relative to other major manufacturing countries according to various metrics. Not all countries compile information on each subject, so it is not possible to show data for the same set of countries on each chart. This report draws on data from a number of sources, and has certain unavoidable statistical problems of which the reader should be aware.

Despite meaningful progress in standardization, countries define "manufacturing" in different ways; in particular, some associate manufacturing with factory production, while others may label a self-employed artisan as a manufacturing worker. Some countries have sophisticated sampling systems to collect data about production and employment from firms and households, whereas others rely heavily on estimates drawn from macroeconomic models or collect data only from a non-random subset of enterprises, such as those located in major cities. International comparisons of compensation data are especially difficult because of national differences in taxation and employee benefits. Complicating matters further, the organizations that compile statistics obtained from national governments may adjust the raw data in different ways to improve compatibility, such that certain figures used to prepare this report may not be identical to those published by national statistical services.

Additionally, analysis of trends in manufacturing is complicated by often arbitrary distinctions between manufacturing and non-manufacturing activity. If, for example, a manufacturing firm owns the trucks that deliver its goods to customers, statisticians will count the truck drivers as manufacturing-sector workers, and their wages will be included in the manufacturing sector's value added. If, however, the manufacturer instead contracts with a separate trucking company to deliver its goods, statisticians will consider the truck drivers to be transport-sector workers and their wages will be included in transport-sector value added, making the manufacturing sector appear smaller—even though there has been no change in the total amount of labor or the tasks performed.

All of these factors argue for caution in the use of these data, and warn against unwarranted assumptions of precision.

How the U.S. Manufacturing Sector Ranks

The standard measure of the size of a nation's manufacturing sector is not manufacturers' sales, but rather their value added. Value added attempts to capture the economic contribution of manufacturers in designing, processing, and marketing the products they sell.

At the level of an individual firm, value added can be calculated as total sales less the total value of purchased inputs, such as raw materials and electricity. The intuition behind this calculation is that a firm that purchases raw materials and processes them only slightly may have substantial sales, but its manufacturing efforts will not have transformed the materials in ways that significantly increase their value. Alternatively, a firm's value added can be measured as the sum of its employee compensation, business taxes (less subsidies), and profits.

The aggregate value added of a country's manufacturing sector cannot be determined simply by adding up the value added of its manufacturers. If a domestic manufacturer uses inputs from its plants abroad, those inputs contain value added by the firm, but not within the United States. Calculating total value added in manufacturing thus requires adjustments for imported parts and components incorporated into the output of domestic factories, and also for domestic products that were exported and used in a foreign plant to make products that were subsequently imported into the United States.[1]

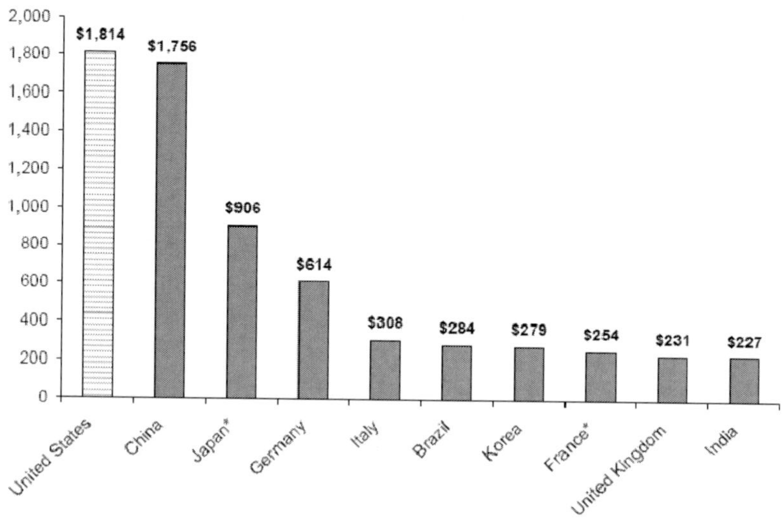

Source: World Bank, http://data.worldbank.org/indicator/NV.IND.MANF.CD.
Note: * Data for Japan and France are for 2009.

Figure 1. Countries with Largest Manufacturing Sectors by Value Added, (Billions of U.S. dollars, 2010).

According to World Bank estimates, the United States retained its position as the largest manufacturing nation in 2010, with value added of $1.8 trillion, closely followed by China. No 2010 data are available for Japan, the third-ranking country, but based on 2009 Japanese data the manufacturing sectors in the United States and China are both roughly twice the size of Japan's (see **Figure 1**). Germany is the only other country whose manufacturing sector is more than one-fifth the size of those in the United States and China.[2]

The U.S. share of global manufacturing value added has declined over time, from nearly one-third in the early 1980s to just short of one-fifth today (see **Figure 2**). Similarly, Japan's share of global manufacturing value added has contracted from 22% in 1993 to around 10% now, and Germany's has fallen from 10% to 6%. These smaller shares are a consequence of the very rapid increase in manufacturing activity in emerging economies, notably China, and do not indicate absolute declines in manufacturing value added in those countries. Manufacturing value added in the United States, as measured by the Bureau of Economic Analysis in inflation-adjusted 2005 dollars, rose 75% from 1990 to 2010 and 16% from 2000 to 2010, although it was lower in 2010 than at the onset of the most recent recession in 2007.

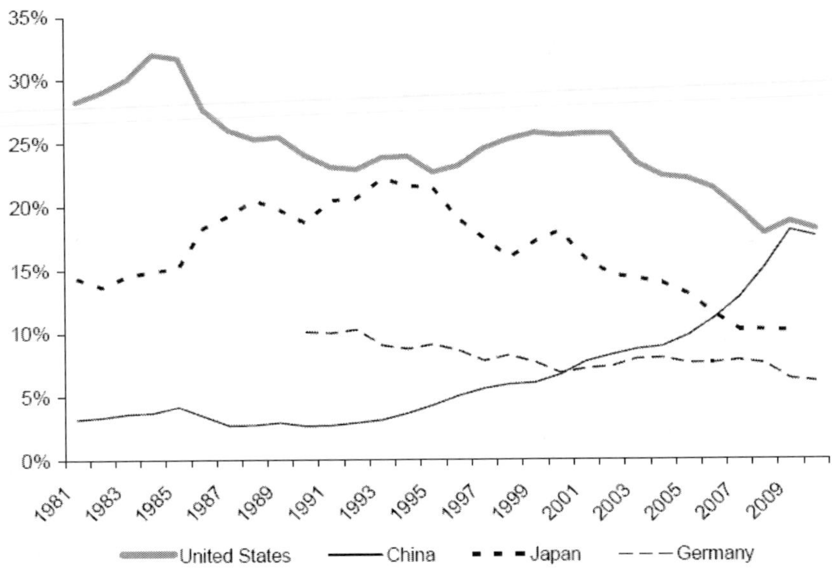

Source: World Bank, http://data.worldbank.org/indicator/NV.IND.MANF.CD.

Figure 2. Selected Countries' Shares of Global Manufacturing Value Added.

Manufacturing value added amounted to 12.4% of total U.S. gross domestic product (GDP) in 2010, according to World Bank estimates. Manufacturing is more significant in the United States, relative to the size of the economy, than in the United Kingdom and France, but much less important than in Japan, Germany, Indonesia, Korea, and China (see **Figure 3**). Chinese manufacturing value added accounted for 29.6% of its economy's total output in 2010, according to the World Bank.

In this respect, it is important to note that a high ratio of manufacturing value added to GDP is not necessarily a sign of economic vibrancy. To the contrary, a high ratio may indicate that various policies or practices, such as labor regulations, credit subsidies, or protection from imports, are standing in the way of a reallocation of capital and labor from manufacturing to other sectors in which they might contribute more to economic growth.

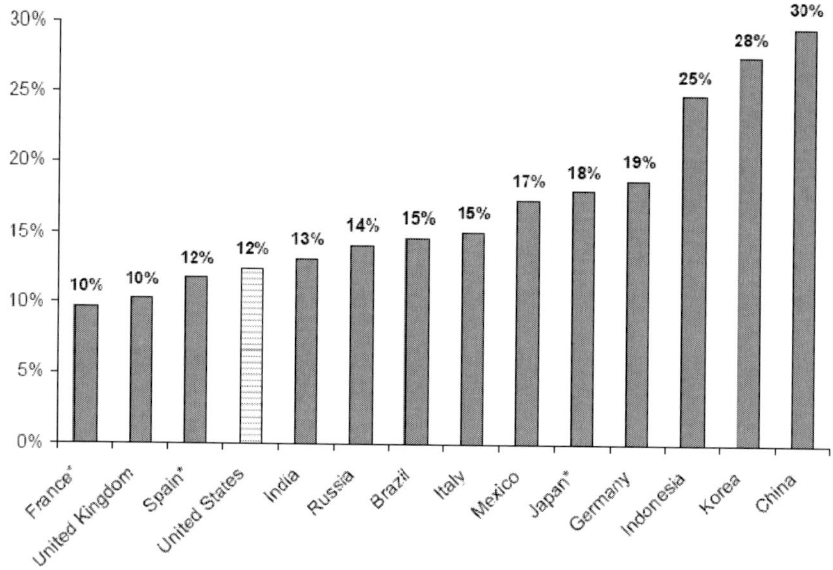

Source: World Bank, computed from data available at http://data.worldbank.org/indicator/NY.GDP.MKTP.CN and http://data.worldbank.org/indicator/NV.IND.MANF.CN.

Note: * Data for France, Japan, and Spain are for 2009. Figures are rounded to nearest percentage point.

Figure 3. Share of Manufacturing in National Economies, (Manufacturing value added as percent of gross domestic product, 2010).

Despite its relatively low rank in manufacturing as a share of GDP, the United States appears to have outperformed most other wealthy countries in the growth of manufacturing value added over the past decade. U.S. value added in manufacturing, adjusted for inflation, rose 11 percent between 2000 and 2010, according to estimates by the U.S. Bureau of Labor Statistics (BLS). Japan and Germany had lower growth in manufacturing value added during that period, after adjusting for inflation, while France, the United Kingdom, Italy, and Canada saw declines in value added.[3] Separate data from the World Bank show that China and Taiwan had much faster growth in value added than the United States, after adjusting for inflation (see **Figure 4**).[4]

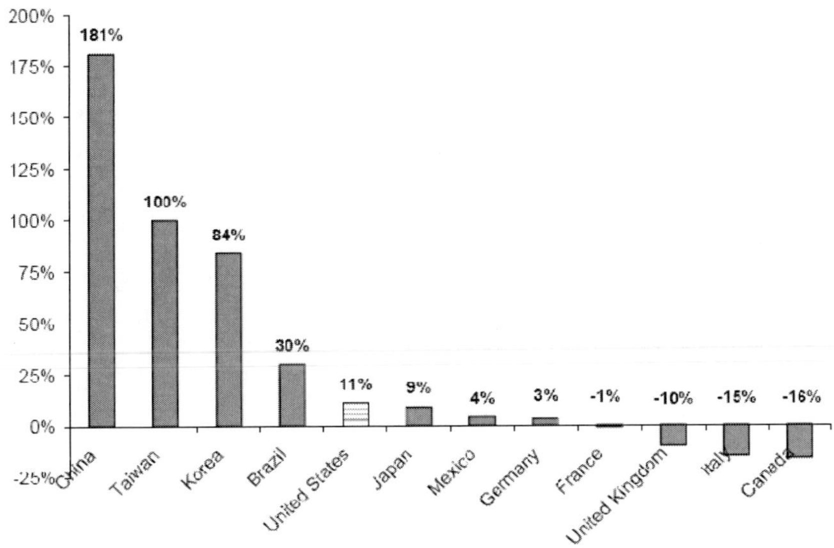

Sources: Derived from U.S. Bureau of Labor Statistics, "International Comparisons of Manufacturing Productivity and Unit Labor Cost Trends: Underlying Data Tables," October 13, 2011. Figures for China, Mexico, and Taiwan derived from World Bank, http://data.worldbank.org/indicator/NV.IND.MANF.KN.

Note: Data for France include mining.

Figure 4. Change in Value Added in Manufacturing, 2000-2010, (Adjusted for inflation in each respective country).

The United States has also performed well in manufacturing, compared to other high-income economies, when viewed over a longer time period. From

1990 through 2010, the only high-income countries with faster growth in manufacturing value added were Finland and Sweden.

Additionally, data on inflows of foreign investment suggest that the United States has been an attractive manufacturing location relative to other high-income countries in recent years. Over the 2007-2009 period, 34.6% of foreign direct investment coming into the United States went into the manufacturing sector, compared to 21.1% in Italy, 18% in the United Kingdom, 11.4% in France and Japan, and less than 10% in Germany and Korea.[5] Comparative data are not available regarding the extent to which foreign direct investment finances construction of new manufacturing facilities as opposed to acquisition of existing facilities.

Data on capital investment in manufacturing are compiled by the Organisation for Economic Cooperation and Development (OECD), a group of 34 nations, most with relatively high per-capita incomes. Investment data are available for only a few countries. These indicate that gross investment in fixed manufacturing capital, such as factories and equipment, accounts for a lower share of GDP in the United States than in the other wealthy countries for which data are available (see **Figure 5**). Gross fixed capital formation across the entire economy is lower relative to GDP in the United States than in most of these countries,[6] but the United States also devotes a smaller share of gross fixed capital formation to manufacturing than the other countries, with the exception of France.[7]

Interpreting the comparative data on investment in manufacturing is problematic. A high ratio of gross fixed capital formation to output is not necessarily positive from an economic point of view; if such investment is generating a low return, then high capital investment could indicate inefficient use of capital. The relatively low level of gross investment in the United States might therefore indicate that U.S. manufacturers pay greater attention to return on capital than their counterparts in other countries. Another explanation might be that U.S. manufacturers face comparatively few obstacles to contracting fabrication or assembly work to manufacturers abroad, whereas other nations may have policies in place to promote domestic fabrication and assembly or to discourage foreign sourcing. Also, it is important to note that the definition of gross fixed capital used by the OECD appears to exclude software, which may represent a greater share of investment by U.S. manufacturers than by those in other countries.[8]

Gross fixed-capital formation in manufacturing measures only additions to the capital stock less subtractions for capital stock destroyed or sold into other sectors of the economy, with no adjustment for the depreciation of the existing capital stock. It therefore does not measure the extent to which the capital stock used in manufacturing changes from year to year. According to the OECD, the *net* stock of fixed capital in U.S. manufacturing rose 3% from 2005 to 2008, after adjusting for inflation, but no comparable data are available for other major manufacturing countries.[9]

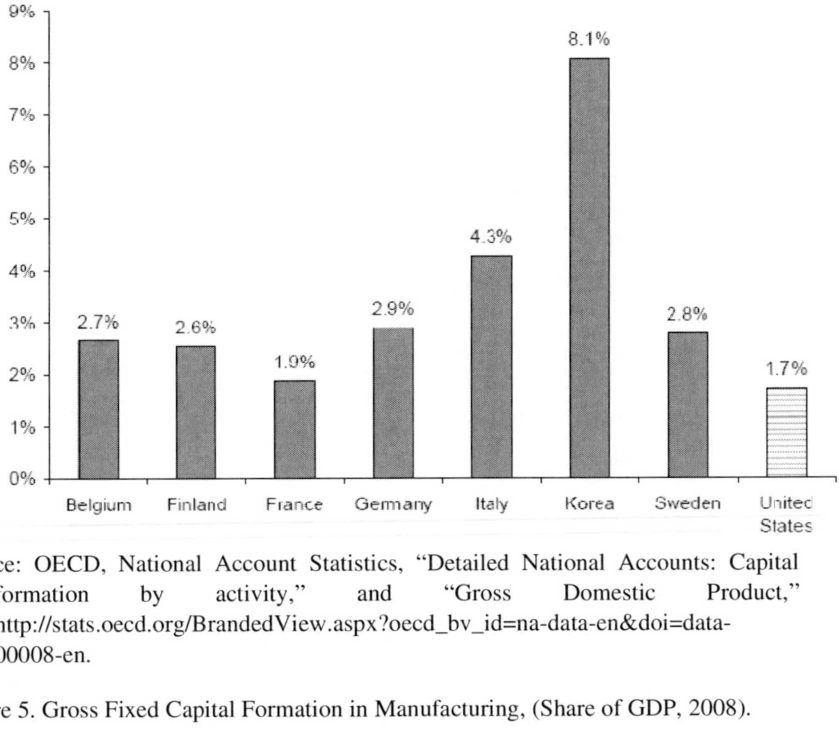

Source: OECD, National Account Statistics, "Detailed National Accounts: Capital formation by activity," and "Gross Domestic Product," http://stats.oecd.org/BrandedView.aspx?oecd_bv_id=na-data-en&doi=data-00008-en.

Figure 5. Gross Fixed Capital Formation in Manufacturing, (Share of GDP, 2008).

The Role of Services in Manufacturing

Measuring manufacturing activity is not without problems, largely because of the imperfect line between manufacturing and services. U.S. statistical agencies, for example, consider activities occurring at establishments whose principal business is manufacturing to be manufacturing, regardless of the specific tasks involved. Similarly, activities occurring at establishments whose principal business is services are considered service activities.

The following three examples will illustrate the statistical confusion that can result. If a manufacturing facility designs and then fabricates a product, the design activities generally count as value added in manufacturing and the workers engaged will be tabulated as manufacturing employees. If the design is created within the manufacturing firm but at a location where no physical production occurs, it could conceivably count as either a manufactured product or a service-sector product. If the manufacturer purchases the design from a specialist design firm, the value added in the design process will be credited to the service sector, and the workers involved will be considered service-sector employees. In all three cases, total employment and total value added are identical; all that differs is the economic sector to which the employment and value added are attributed.

Efforts to measure the value of manufacturing-related services more accurately are still in their infancy. Such data as are available indicate that service-sector inputs incorporated into manufactured products account for a larger share of manufacturing value added in the United States than in any other major economy (see **Figure 6**). Further, the service-sector share of the total value added of manufactured goods increased faster in the United States than in any of the 37 other countries studied between 1995 and 2005.

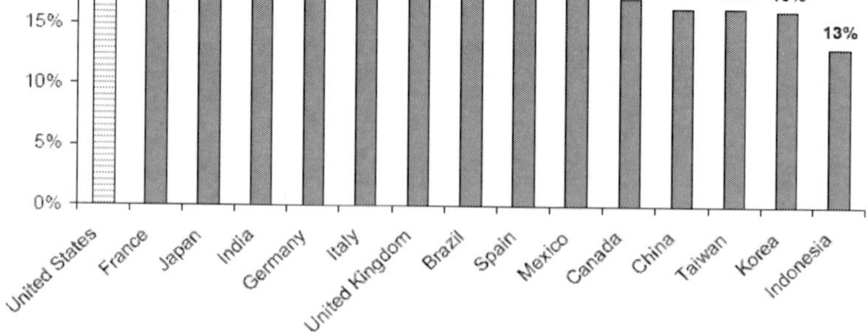

Source: Organisation for Economic Co-operation and Development (OECD), STAN Input-Output Database, May 2011, http://dx.doi.org/10.1787/888932487628.

Figure 6. Service-Sector Inputs into Manufacturing, (Service-sector value added in manufactured goods as percentage of total value added of manufactured goods, 2005).

The figures illustrated in **Figure 6** show only the importance of services purchased by manufacturers from outside firms. One possible interpretation of these data is that U.S. manufacturers are less vertically integrated than those in other countries, such that they more frequently contract with outside providers for services rather than producing them in-house. However, data on the occupations of manufacturing workers argue against this interpretation. In 2008, more than half of all Americans employed within the manufacturing sector worked in service occupations, such as management, technical support, and sales (see **Figure 7**). This is a far greater proportion than in other OECD economies. The relatively high service-intensity of U.S. manufacturing is thus evident within manufacturing firms as well as in their purchases of inputs from outside firms.

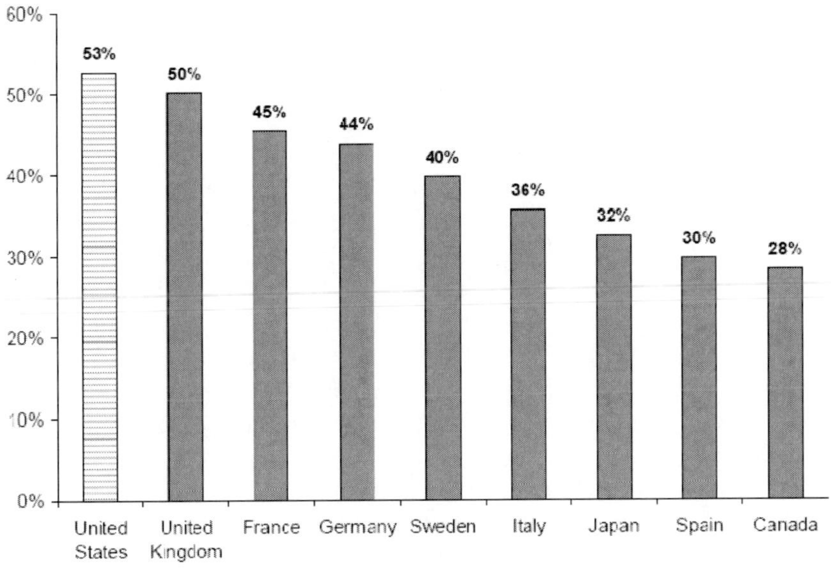

Source: OECD Science, Technology and Industry Scorecard, 2011.

Notes: Swedish data are for 2007. Service-related occupations include (1) legislators, senior officials, and managers; (2) professionals; (3) technicians and associate professionals; (4) clerks; and (5) service workers and shop and market sales workers as defined in the International Standard Classification of Occupations, 1988.

Figure 7. Services-Related Occupations in Manufacturing Industries, (Percentage of all employees in manufacturing, 2008).

In combination, the data in **Figure 6** and **Figure 7** suggest that U.S. manufacturers may be relatively advanced, in comparison to those in other countries, when it comes to automating routine production work, and therefore employ a smaller proportion of their workers in production operations. A related interpretation of these data would be that U.S. manufacturers' output contains a higher proportion of non-physical value, such as intellectual property, than the output of other countries, possibly implying that U.S. manufacturers produce more advanced products. Another possibility is that U.S. manufacturers make greater use of certain services, such as legal, tax, and accounting services, than manufacturers in other countries.

MANUFACTURING WORK

International comparisons of manufacturing employment trends are hampered by inadequate data, particularly for emerging economies. Among the top-ranking manufacturing countries, China, Brazil, and India do not report complete information on manufacturing employment at the national level. Mexico has a nationwide statistical sampling program, but due to definitional and methodological changes a consistent time series is available only since 2009.[10] Manufacturing employment in the United States fell by roughly one-third from 2000 through 2010. Among the major manufacturing countries for which data are available, only the United Kingdom saw a larger decline in manufacturing employment over that period (see **Figure 8**). Looked at over a 20-year period, however, manufacturing employment fell by approximately the same percentage in the United States as in France, Germany, and Japan, and much less than in the United Kingdom (see **Figure 9**). These figures indicate that the diminished importance of manufacturing as a source of jobs is a widespread phenomenon and is not limited to the United States.[11] However, the timing of decline has differed from country to country. Some countries experienced sharp manufacturing employment declines during the 1990s. In other countries, including the United States and Canada, employment in manufacturing held fairly steady during the 1990s but declined quite steeply from 2000 to 2010. The international comparison of manufacturing employment is somewhat different if viewed in terms of hours worked rather than by the number of workers. By this metric, the United States and the United Kingdom had approximately the same rates of decline between 2000 and 2010, far beyond that in other countries (see **Figure 10**). Over the past two decades, however, France, Japan, and the United Kingdom all had larger declines in hours worked in manufacturing than did the United States, and Germany was on a par with the United States (see **Figure 11**). As with manufacturing employment, the timing of declines in hours worked in manufacturing has varied among countries. France, Germany, Italy, Japan, and Korea all had much sharper declines in manufacturing hours worked than did the United States between 1990 and 2000, but the United States experienced a far more rapid decline than those countries between 2000 and 2010.

Whether the measure in the number of workers employed in the sector or the number of work hours, the United States is not unique in experiencing a decline in the need for labor in the manufacturing sector. Even in Korea and Taiwan, where manufacturing output has expanded far more rapidly than in the United States, factories require fewer total hours of labor than was the case a decade ago.

The reduced demand for labor is directly related to improved labor productivity in manufacturing. Manufacturing labor productivity increased much more rapidly in the United States between 2000 and 2010 than in Canada, European countries, or Japan, as measured by real output per hour of manufacturing labor (see **Figure 12**). Taiwan and Korea both had greater improvement in manufacturing labor productivity than the United States.

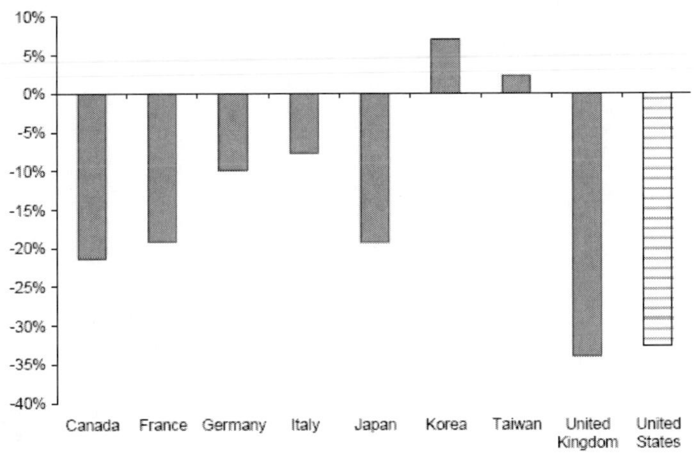

Source: Derived from U.S. Bureau of Labor Statistics, "International Comparisons of Manufacturing Productivity and Unit Labor Cost Trends: Underlying Data Tables," October 13, 2011.

Note: Data for France include mining.

Figure 8. Manufacturing Employment, (Percentage change, 2000-2010).

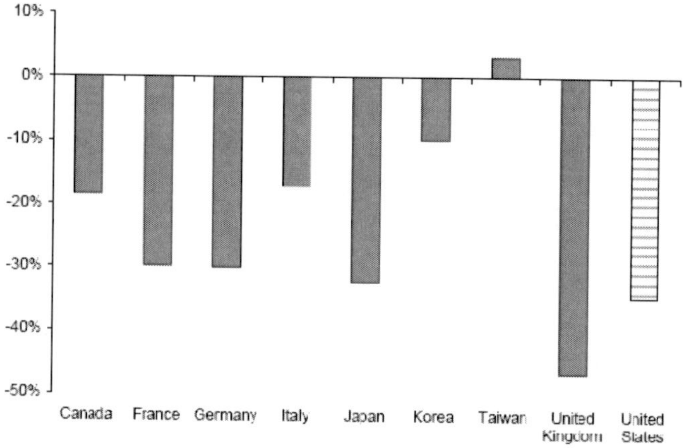

Source: Derived from U.S. Bureau of Labor Statistics, "International Comparisons of Manufacturing Productivity and Unit Labor Cost Trends: Underlying Data Tables," October 13, 2011.

Note: Data for France include mining.

Figure 9. Manufacturing Employment, (Percentage change, 1990-2010).

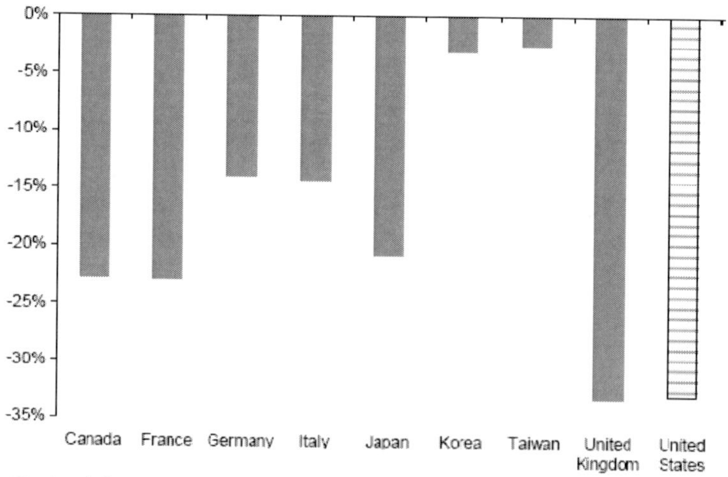

Source: Derived from U.S. Bureau of Labor Statistics, "International Comparisons of Manufacturing Productivity and Unit Labor Cost Trends: Underlying Data Tables," October 13, 2011.

Note: Data for France include mining.

Figure 10. Manufacturing Hours Worked (Percentage change, 2000-2010).

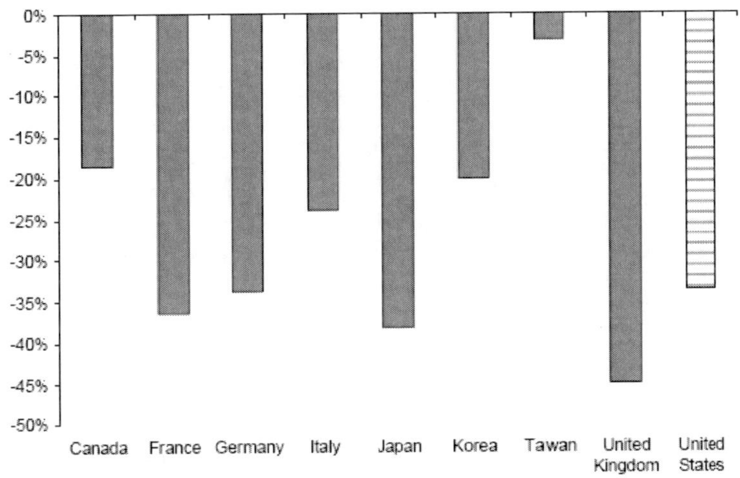

Source: Derived from U.S. Bureau of Labor Statistics, "International Comparisons of Manufacturing Productivity and Unit Labor Cost Trends: Underlying Data Tables," October 13, 2011.
Note: Data for France include mining.

Figure 11. Manufacturing Hours Worked, (Percentage change, 1990-2010).

The strong improvement in U.S. labor productivity in manufacturing has several causes. One is manufacturers' large investments in automation, which have eliminated many routine assembly jobs; less than 40% of the workers in U.S. manufacturing establishments are now directly engaged in production. A related factor is the rapid increase in education levels among U.S. manufacturing workers, some 28% of whom possess college degrees.[12] A third cause of improvement in average manufacturing productivity is the relatively rapid growth of certain areas of U.S. manufacturing in which labor productivity is extremely high. These include instruments manufacturing, in which output grew 28% from 2000 to 2010, and aerospace manufacturing, which expanded output 21% over the same period, even as total U.S. manufacturing output fell 3%.[13] In part, however, the measured improvement in labor productivity in manufacturing also reflects the rapid shrinkage of low-productivity manufacturing activities over the course of the past decade. During this period, many manufacturers moved routine assembly work abroad, either to their own factories or to those of contract suppliers. The reduction of U.S. import barriers encouraged apparel imports and the resulting reduction of domestic capacity in the low-productivity apparel industry. As U.S. plants with below-average productivity closed, the average labor productivity of the remaining manufacturing plants necessarily increased even in the absence of productivity improvements.[14] Similarly, the very rapid increases in manufacturing labor productivity measures in Korea and Taiwan likely reflect the closure of low-productivity manufacturing as well as the expansion of capital-intensive manufacturing and rising education levels among manufacturing workers. For example, Korea's exports of apparel, the product of a comparatively low-productivity industry, declined from $5 billion in 2000 to $1.6 billion in 2010, and Taiwan's fell from $3 billion to $1 billion over the same period.[15] As the jobs involved in producing such goods were eliminated, the average productivity of those countries' manufacturing workers would have risen even without growth in high-productivity sectors. At the other extreme, Italy, which saw only a small drop in manufacturing employment over the decade, registered absolute declines in manufacturing value added and in output per hour worked. This may indicate that restructuring low-productivity operations has been a challenge for Italian manufacturers, or may be the result of other government policies to retain employment. Italy's export data provide some evidence of this: after adjusting for the appreciation of the euro, Italy's exports of apparel rose 9% between 2000 and 2010.[16]

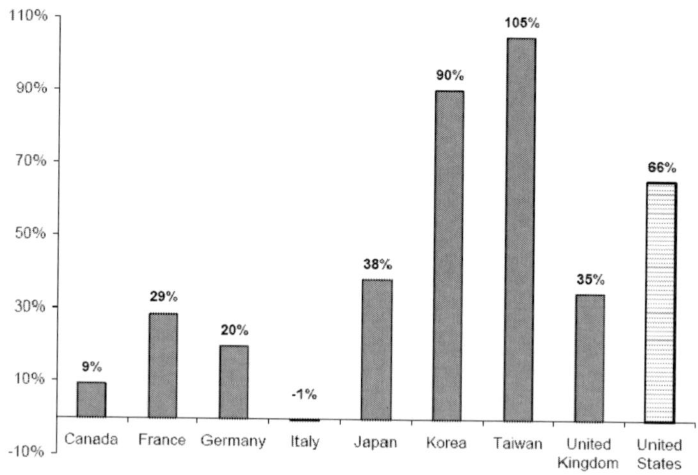

Source: U.S. Bureau of Labor Statistics, "International Comparisons of Manufacturing Productivity and Unit Labor Cost Trends: 2010 Data Tables," December 1, 2011.

Figure 12. Real Output per Labor Hour in Manufacturing, (Percentage change, 2000-2010).

Average compensation per employee in U.S. manufacturing was $34.74 per hour in 2010, a 36% increase since 2000.[17] U.S. hourly manufacturing labor costs were lower than those in 13 of 33 countries studied by the Bureau of Labor Statistics (BLS). Due in good part to exchange-rate changes, average compensation per hour expressed in U.S. dollar terms has been rising more slowly in the United States than in most other major manufacturing countries (see **Table 1**). For example, hourly compensation in manufacturing in Brazil rose an average of 6.7% annually between 1997 and 2010, compared to a 3.2% annual rate of increase in the United States, but after exchange-rate changes are factored in, Brazilian manufacturing labor costs declined relative to those in the United States.

Table 1. Hourly Compensation Costs in Manufacturing (U.S. dollar basis)

	Direct Pay, 2010	Total Compensation Costs, 2010	Average Annual Percentage Change, 1997-2010
Brazil	$6.85	$10.08	2.8%
Canada	$27.64	$35.67	5.0%
France	$27.61	$40.55	3.8%
Germany	$34.24	$43.76	3.2%
Italy	$23.84	$33.41	4.2%
Japan	$26.29	$31.99	2.8%
Korea	$13.36	$16.62	4.5%
Mexico	$4.30	$6.23	4.6%
Taiwan	$7.13	$8.36	1.3%
United Kingdom	$25.05	$29.44	3.6%
United States	$26.27	$34.74	3.2%

Source: U.S. Bureau of Labor Statistics, "International Comparisons of Hourly Compensation Costs in Manufacturing, 2010," December 21, 2011, Tables 2 and 3.

Notes: "Direct Pay" includes vacation pay, bonus payments, and employer contributions to employees' savings funds. "Total Compensation Costs" includes pensions, disability insurance, sick leave, health insurance, severance pay, other social insurance expenditures, and taxes on payrolls or employment. "Average Annual Percentage Change" is calculated in terms of U.S. dollars and incorporates the effects of exchange-rate changes.

Accurate nationwide data on manufacturing compensation costs in China and India are not available. BLS estimates average manufacturing compensation in China to have been $1.36 per hour in 2008,[18] but it warns that this estimate is not as robust as those for other countries.[19] With respect to

India, BLS estimates average compensation in formal manufacturing establishments to have been $1.17 in 2007, but cautions that this figure overstates average compensation as it pertains to only about 20% of the country's manufacturing workers.[20] Because data from China and India are not comparable to those from other countries, they are not included in **Table 1**.

The data on average hourly compensation costs can be misleading, as they are not adjusted for differences in the industrial mix. In most countries, including the United States, labor costs vary greatly among industries; the average hourly wage of U.S. workers who make household furniture is less than $16 per hour, whereas average hourly wage in aircraft manufacturing exceeds $40.

The most recent U.S. government data on comparative compensation costs within individual industries, which date to 2007, show U.S. costs to be lower than those in major European countries, although well above those in emerging economies (see **Table 2**).

The more detailed data that would be required to correct for national differences in the products manufactured by these industries are not available.

Table 2. Hourly Compensation Costs in Selected Manufacturing Industries (U.S. dollar basis, 2007)

	Wood Products	Textiles	Chemicals	Machinery	Motor Vehicles
Brazil	$3.64	$4.58	$13.89	$8.78	$13.08
Canada	$27.92	$23.53	$37.10	$34.06	$41.11
France	$28.13	$28.66	$50.81	$39.09	$42.73
Germany	$35.11	$36.94	$66.76	$54.17	$67.14
Italy	$24.57	$29.23	$44.87	$34.31	$33.36
Japan	$17.49	$19.62	$34.59	$26.78	NA
Korea	$13.80	$12.02	$24.07	$18.42	$21.66
Mexico	$2.06	$3.04	$9.06	$4.65	$5.29
Taiwan	$5.78	$6.66	$10.91	$7.38	$9.08
United Kingdom	$26.66	$29.68	$46.93	$37.57	$41.65
United States	$20.67	$21.17	$45.76	$32.08	$33.77

Source: U.S. Bureau of Labor Statistics, ftp://ftp.bls.gov/pub/special.requests/ForeignLabor/aeindustrynaics.txt.

Note: Industry definitions are not identical in all countries.

TECHNOLOGY AND RESEARCH IN MANUFACTURING

High-technology manufacturing has been a particular focus of public-policy concern for many years. There is no standard definition of high-tech manufacturing, but commentators have long asserted that high-technology production has especially beneficial economic spillovers.[21] Although definitions of "high-tech industry" vary, the OECD considers that manufacturing of pharmaceuticals; office, accounting, and computing machinery; radio, television, and communications equipment; medical, precision, and optical instruments; and aircraft and spacecraft is particularly technology-intensive, based on those industries' research and development (R&D) expenditures and on the amount of R&D embodied in their products.[22] It is important to note in this context that some industries that may have a considerable technological component, such as automobile and machinery manufacturing, are not considered high-technology industries by the OECD. The United States derives a greater share of manufacturing value added from high-tech industries than is the case in most other OECD member countries (see Figure 13). Moreover, the share of value added represented by high-technology sectors has been rising in the United States, whereas it has been stable or declining in many other countries.

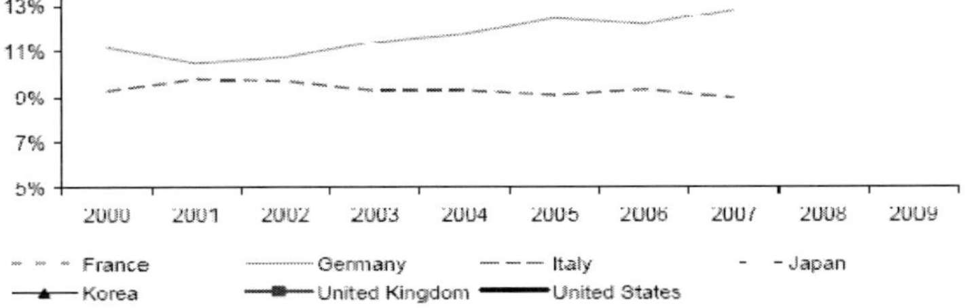

Source: OECD STAN database, http://stats.oecd.org/Index.aspx?DatasetCode-STAN08BIS&lang=en.

Figure 13. Importance of High-Tech Industries, (Share of country's manufacturing value added).

Manufacturers in the United States spend far more on research than those in any other major industrial country. Adjusting for differences in purchasing power, spending on manufacturing research and development was nearly twice as high in the United States as in Japan in 2007, and almost four times the level of Germany (see **Figure 14**).[23]

Although far less manufacturing R&D occurs in countries that have industrialized more recently, R&D spending in those countries has been growing at a very rapid rate (see **Figure 15**).

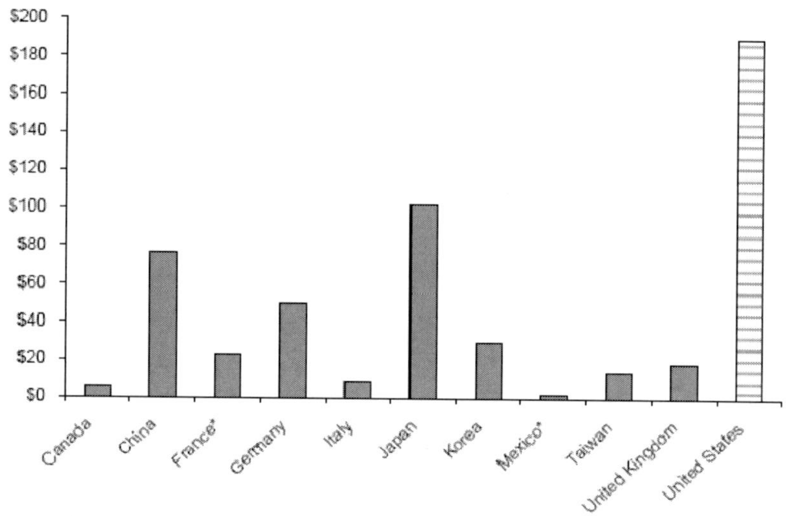

Source: OECD STAN database, "STAN R&D expenditures in Industry," http://stats.oecd.org/index.aspx.

Note: * Mexico data are for 2007.

Figure 14. R&D in Manufacturing, 2008, (Billions of U.S. dollars at purchasing power parity).

Manufacturers have been responsible for approximately 70% of all R&D conducted by businesses in the United States in recent years. This is similar to the proportion in Italy, but far lower than in Germany, Japan, and Korea, where manufacturers account for 88%, on average, of all business-financed R&D. Conversely, the service sector is relatively more important in undertaking research and development in the United States than in many other countries. The most notable exception is the United Kingdom, where service companies account for three-fifths of all business R&D spending.[24]

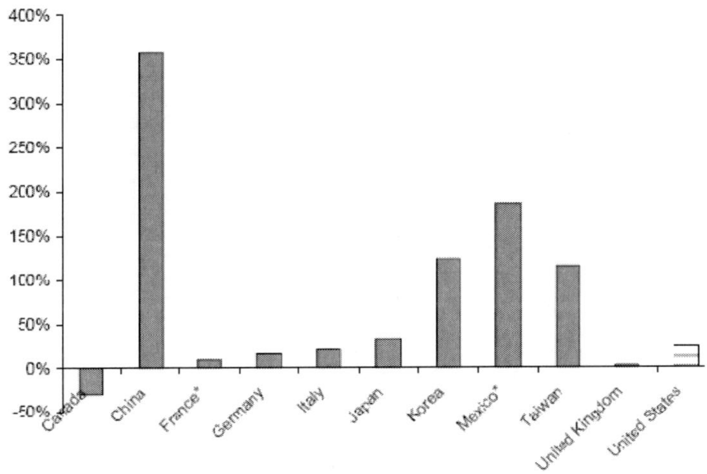

Source: OECD STAN database, "STAN R&D expenditures in Industry," http://stats.oecd.org/index.aspx.

Note: * Mexico data are for 2000-2007.

Figure 15. Growth in Manufacturing R&D, (Change in real local currency, 2000-2008).

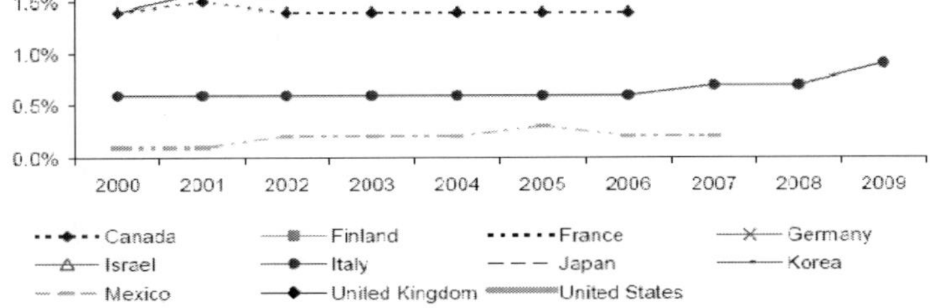

Source: OECD, STAN indicators: R&D intensity of manufacturing sectors, http://www.oecd-ilibrary.org/industry-and-services/data/stan-oecd-structural-analysis-statistics/stan-indicators-2009_data-00031-en.

Figure 16. Manufacturers' Research Intensity in Selected Countries.

The research intensity of U.S. manufacturing increased during the first decade of the twenty-first century, indicating that U.S. manufacturers are devoting a growing share of their revenue to R&D. In 2000, U.S. manufacturers spent 2.9% of sales on research and development, a figure that rose to 3.3% by 2008. The only country in which manufacturers' R&D spending has been growing at a faster rate is Korea. U.S. manufacturers devote a greater proportion of their revenue to R&D than those in any other country save Japan, including some countries renowned for their relatively large high-technology sectors, such as Finland and Israel (see **Figure 16**).[25]

One possible reason for national differences in R&D intensity in manufacturing is differences in the composition of the manufacturing sector. Industries such as aircraft and spacecraft manufacturing and electronic instrument manufacturing are among the most research-intensive in every country, and, all other things equal, countries in which these sectors are relatively large may be expected to have greater R&D intensity in manufacturing than countries in which they are less important.

Table 3 provides an alternative cross-country comparison of R&D spending by manufacturers by breaking out R&D intensity by industry. The data pertain to 2006, the most recent year for which comparable data are available for all countries shown. The table illustrates the fact that manufacturers in the United States are more research-intensive than those in other countries only in selected industries, such as electronic instruments. In other industries, foreign manufacturers spend comparatively more on R&D than those in the United States. For example, Japanese manufacturers of office, accounting, and computing machinery devote a greater share of sales to R&D than those in any other country, and Italy, whose manufacturers generally are much less R&D-intensive than those in other countries, appears to have particularly extensive industry research related to aerospace manufacturing.

Table 4 confirms the implication of **Table 3** that manufacturers' R&D spending is targeted quite differently in different countries. In the United States, a much larger proportion of manufacturing R&D occurs in the pharmaceutical sector than is the case elsewhere, with the exception of the United Kingdom. The instruments sector, including medical equipment and process-control equipment as well as navigational, testing, and measuring equipment, is also disproportionately important in the United States. By contrast, the motor vehicle sector accounts for a significantly smaller share of manufacturers' research and development activity in the United States than in other countries for which data are available.

Table 3. Comparative Research and Development Spending by Industry (R&D outlays by manufacturers as a percentage of sales, 2006)

	Canada	France	Germany	Italy	Japan	Korea	United Kingdom	United States
All manufacturing	1.4%	2.5%	2.4%	0.6%	3.7%	1.9%	2.4%	3.3%
Pharmaceuticals	11.9%	8.7%	10.4%	1.5%	15.0%	2.5%	24.9%	22.5%
Office, accounting, computing machinery	10.9%	7.9%	4.1%	1.1%	28.7%	3.9%	0.4%	11.0%
Electrical machinery	1.3%	3.5%	1.3%	0.5%	8.8%	1.4%	3.3%	2.0%
Instruments	NA	7.1%	6.6%	2.4%	14.4%	2.2%	3.6%	18.0%
Motor vehicles	0.5%	4.7%	4.4%	1.8%	4.3%	2.8%	1.9%	3.4%
Aircraft and spacecraft	6.3%	5.2%	10.4%	12.5%	4.2%	9.0%	10.7%	11.3%

Source: OECD, STAN indicators: R&D intensity of manufacturing sectors, http://www.oecd-ilibrary.org/industry-and-services/data/stan-oecd-structural-analysis-statistics/stan-indicators-2009_data-00031-en.

Table 4. Manufacturers' R&D Spending by Sector (Percentage of total research and development spending by manufacturers)

Country	Year	Pharma	Telecoms	Instruments	Motor Vehicles	Other Trans. Equ.	Other
France	2007	6.9%	14.8%	9.7%	11.0%	19.6%	38.1%
Germany	2008	8.3%	8.0%	8.0%	36.9%	5.7%	33.1%
Italy	2008	6.9%	10.9%	6.9%	16.0%	15.6%	43.6%
Japan	2008	10.9%	20.5%	5.8%	19.3%	0.5%	42.9%
Korea	2008	2.8%	51.6%	3.0%	15.0%	2.4%	25.3%
United Kingdom	2008	N/A	5.6%	11.1%	17.1%	23.6%	N/A%
United States	2006	22.8%	18.1%	13.1%	9.7%	11.7%	24.7%

Source: OECD, Science, Technology and R&D Statistics, "Business enterprise R-D expenditure by industry," http://www.oecd-ilibrary.org/content/data/data-00183-en.

The United States ranks third among OECD member countries, following only Ireland and Finland, in the proportion of manufacturing R&D that occurs in high-technology sectors. In the United States, OECD reports, 69% of manufacturers' total R&D spending in 2007 occurred in high-technology sectors and 22% in medium-technology sectors. In Germany, by contrast, 60% of manufacturers' R&D spending occurred in medium-technology sectors, such as motor vehicle and machinery manufacturing, and the corresponding figure for Japan was 45%.[26]

End Notes

[1] For more on the changing nature of value added in manufacturing, see CRS Report R41712, *"Hollowing Out" in U.S. Manufacturing: Analysis and Issues for Congress*, by Marc Levinson.

[2] See http://data.worldbank.org/indicator/NV.IND.MANF.CD. The data used here are standardized and hence may differ from those produced by national statistical services; for example, the World Bank estimates U.S. manufacturing value added at $1.814 trillion in 2010, whereas the U.S. Bureau of Economic Analysis has published an estimate of $1.702 trillion; see http://www.bea.gov/iTable/iTable.cfm?ReqID=5&step=1. The U.S. Census Bureau, which uses a different method of calculation, gives 2010 manufacturing value added as $2.19 trillion; see *2010 Annual Survey of Manufactures*, item AM1031AS101, http://factfinder2.census.gov/faces/nav/jsf/pages/index.xhtml.

[3] U.S. Bureau of Labor Statistics, "International Comparisons of Manufacturing Productivity and Unit Labor Cost Trends: Underlying Data Tables," October 13, 2011. The BLS estimates of change in real value added cited in this paragraph differ from the figures presented by the Bureau of Economic Analysis, cited above, as BLS has made adjustments for international compatibility.

[4] http://data.worldbank.org/indicator/NV.IND.MANF.KN.

[5] OECD International Direct Investment Statistics, "Foreign direct investment: flows by industry," http://doi:10.1787/ data-00334-en.

[6] http://stats.oecd.org, "National Accounts at a Glance: 6. Capital," indicator K1S: Consumption of fixed capital, percentage of GDP.

[7] Some 9.6% of U.S. fixed-capital formation in 2008 occurred in the manufacturing sector, compared to 8.5% in France. The highest proportion among the countries for which data are available was 27.5% in Korea. See OECD, "Detailed National Accounts: Capital formation by activity," *OECD National Accounts Statistics* (database), http://doi: 10.1787/data-00009-en.

[8] OECD uses the definition established by the United Nations Statistics Division, which reads: "Gross fixed capital formation is measured by the total value of a producer's acquisitions, less disposals, of fixed assets during the accounting period plus certain additions to the value of non-produced assets (such as subsoil assets or major improvements in the quantity, quality or productivity of land) realised by the productive activity of institutional units." http://unstats.un.org/unsd/snaama/glossresults.asp?gID=34.

[9] OECD, "Detailed National Accounts: Fixed assets by activity and by type of product," *OECD National Accounts Statistics* (database), http://doi: 10.1787/data-00009-en.

[10] On manufacturing employment in China, see Judith Banister and George Cook, "China's employment and compensation costs in manufacturing through 2008," *Monthly Labor Review*, March 2011, p. 39, http://www.bls.gov/ opub/mlr/2011/03/art4full.pdf. On manufacturing employment in India, see Jessica R. Sincavage, Carl Haub, and O.P. Sharma, "Labor costs in India's organized manufacturing sector," *Monthly Labor Review*, May 2010, p. 3, http://www.bls.gov/opub/mlr/2010/05/art1full.pdf. Recent Mexican data from the

Instituto Nacional de Estadística y Geografía are available at http://dgcnesyp.inegi.org.mx /cgi-win/bdiecoy.exe/445?s=est&c=25534.

[11] These data are compiled by the U.S. Bureau of Labor Statistics (BLS) and adjusted for consistency. For most countries, the manufacturing sector is as defined by the International Standard Industrial Classification system, but data for Canada and the United States are in accordance with the North American Industry Classification System. The data for France include some mining activity. For details, see the BLS detailed technical notes available at http://www.bls.gov/fls/intl_prod_tn.pdf.

[12] On occupations and education within the manufacturing sector, see CRS Report R41898, *Job Creation in the Manufacturing Revival*, by Marc Levinson.

[13] Output changes are calculated from annual figures published in the Federal Reserve Board G.17 release, "Industrial Production and Capacity Utilization."

[14] In general, the manufacturing industries with the lowest productivity growth are those in which it has proven most difficult to automate production processes to increase output per worker hour. The apparel and footwear industries are notable in this respect. From 1973 to 2001, multifactor productivity grew at an annual rate of 0.9% for all U.S. manufacturing, but at only 0.7% for apparel and 0.3% for leather and leather products. For detailed data on this subject, see U.S. Bureau of Labor Statistics, "Multifactor Productivity in U.S. Manufacturing and in 20 Manufacturing Industries, 1949-2001," February 10, 2004, http://www.bls.gov/mfp/tables.htm,and "Manufacturing Sector and NIPAlevel Manufacturing Industries KLEMS multifactor Productivity Tables by Measure," August 11, 2011, http://www.bls.gov/mfp/mprdload.htm.

[15] Korea and Taiwan data were taken from World Trade Organization statistics database, http://stat.wto.org/ StatisticalProgram/WSDBViewData.aspx?Language=E.

[16] In current U.S. dollars, Italian apparel exports rose from $13.4 billion in 2000 to $20 billion in 2010, a 49% increase. See http://stat.wto.org/StatisticalProgram/WSDBView Data.aspx? Language=E. Over the same period, the euro appreciated approximately 40% against the dollar.

[17] "Compensation" includes pay for time worked, employee benefits, and labor-related taxes net of subsidies.

[18] See Banister and Cook, "China's employment and compensation costs in manufacturing through 2008."

[19] BLS News Release, "International Comparisons of Hourly Compensation Costs in Manufacturing, 2010," December 21, 2011, http://www.bls.gov/news.release/ichcc.nr0.htm.

[20] Sincavage, et al, "Labor costs in India's organized manufacturing sector."

[21] See, for example, Stephen S. Cohen and John Zysman, *Manufacturing Matters: The Myth of the Post-Industrial Economy* (New York, 1987), p. 106, and Lester Thurow, *Head to Head: The Coming Economic Battle Among Japan, Europe, and America* (New York, 1992), pp. 45-51.

[22] These sectors correspond to United Nations International Standard Industrial Classifications 2423, 30, 32, 33, and 353. For details, see OECD, "ISIC Rev. 3 Technology Intensity Definition," July 7, 2011, p. 1, http://www.oecd.org/ dataoecd/43/41/48350231.pdf.

[23] These figures include expenditures by manufacturers, whatever the original source of the funds. For technical background, see OECD, "The OECD Analytical BERD (ANBERD) Database," August 5, 2011, http://www.oecd.org/ dataoecd/52/23/47840198.pdf.

[24] OECD, Science, Technology and R&D Statistics, "Business enterprise R-D expenditure by industry," http://www.oecd-ilibrary.org/content/data/data-00183-en.

[25] The data discussed in this paragraph measure research and development expenditures by manufacturers as a percentage of their sales. OECD also compiles statistics on the ratio of R&D spending to value added. However, these statistics can be problematic for single-year, cross-country comparisons, as a decline in an industry's profitability can reduce its value added, increasing the ratio of R&D outlays to value added even if no additional research is undertaken.

[26] OECD Science, Technology and Industry Scoreboard 2011, "Business R&D in the manufacturing sector by technological intensity," Figure 6.8.2, http://www.oecd-ilibrary.org/sites/sti_scoreboard-2011-en/06/08/index.html?contentType=/ns/Chapter,/ns/StatisticalPublication&itemId=/content/chapter/sti_scoreboard-2011-62-en& containerItemId=/content/serial/20725345&access ItemIds=&mimeType=text/html.

In: United States Manufacturing
Editors: V. Murphy and J. E. Walker
ISBN: 978-1-62081-522-9

Chapter 2

"HOLLOWING OUT" IN UNITED STATES MANUFACTURING: ANALYSIS AND ISSUES FOR CONGRESS*

Marc Levinson

SUMMARY

The health of the U.S. manufacturing sector has been a long-standing concern of Congress. Although Congress has established a wide variety of tax preferences, direct subsidies, import restraints, and other federal programs with the goal of retaining or recapturing manufacturing jobs, only a small proportion of U.S. workers is now employed in factories. Meanwhile, U.S. factories have stepped up production of goods that require high technological sophistication but relatively little direct labor. Labor productivity in manufacturing, as measured by government data, has grown rapidly, suggesting that the manufacturing sector as a whole remains healthy.

Recent data, however, challenge the belief that the manufacturing sector, taken as a whole, will continue to flourish. Unlike previous expansions, the two most recent cyclical upturns in the U.S. economy have not generated jobs in manufacturing. Moreover, statistics suggest that domestic value represents a diminishing share of the value of U.S. factory output. One interpretation of these data is that manufacturing is

* This is an edited, reformatted and augmented version of a Congressional Research Service publication, CRS Report for Congress R41712, from www.crs.gov, dated March 23, 2011.

"hollowing out" as companies undertake a larger proportion of their high-value work abroad. These developments raise the question of whether the United States will continue to generate highly skilled, high-wage jobs related to advanced manufacturing.

The evidence concerning "hollowing out" is ambiguous, as conceptual issues and statistical deficiencies make it difficult to determine whether the recent decline in manufacturing value added, relative to shipments, is a short-term phenomenon or a long-term trend. Despite improvements in recent years, U.S. statistical agencies still tend to treat manufacturing and services as unrelated economic activities, and it is not clear that existing data series on domestic economic activity, trade, and freight transportation adequately capture changes in the nature of manufacturing, the sources of employment, and the creation of value.

Nonetheless, evidence suggests strongly that physical production activities account for a diminishing share of the final value of manufactured products, with service-related inputs such as research, product design, and marketing becoming more important. Further, the production of many goods is dispersed across multiple locations along global supply chains, making it difficult to determine where value is added. Such shifts pose a challenge to efforts to capture economic value by promoting goods production in the United States.

In the context of national security, the fact that U.S. manufacturers of vital products are critically dependent upon inputs from abroad is frequently a subject of concern. International comparisons indicate that the United States is in no way unique in its dependence on foreign inputs to manufacturing. Although the output of U.S. factories contains a large proportion of foreign value added, many other countries appear to be even more dependent upon foreign value added than is the United States, at least with respect to goods traded in international markets.

INTRODUCTION

The health of the U.S. manufacturing sector has been a major concern of Congress for more than three decades. Over the years, Congress has enacted a wide variety of tax preferences, direct subsidies, import restraints, and other federal programs intended to bolster the manufacturing sector, often with the goal of retaining or recapturing highly paid manufacturing jobs.

Only a small proportion of U.S. workers is now employed in factories, as manufacturers have shifted low-value, labor-intensive production, such as apparel and shoe manufacturing, to other countries. Meanwhile, U.S. factories have stepped up production of goods that require high technological sophistication but relatively little direct labor. Despite highly publicized factory closures, the good-producing capacity of the U.S. economy remains near its all-time peak, as measured by the Federal Reserve Board.

Recent data, however, challenge the belief that the manufacturing sector, taken as a whole, will continue to flourish. In particular, statistics showing that domestic value added represents a diminishing share of the value of U.S. factory output have been interpreted by many analysts as indicating that manufacturing is "hollowing out" as U.S. companies undertake more high-value work abroad. Economic data have been slow to take note of this development, which raises the question of whether the United States will continue to generate highly skilled, high-wage jobs related to advanced manufacturing.

This report discusses economic evidence related to the "hollowing out" thesis with respect to the manufacturing sector. It then considers the policy implications of the debate.

The Health of U.S. Manufacturing

The United States has a very large manufacturing sector. In 2010, manufacturers' shipments reached $5 trillion, equal to one-third of the gross domestic product. Although many factories closed or reduced production during the 2007-2009 recession, output has rebounded since the summer of 2009. In February 2011, the Federal Reserve Board's index of industrial production in manufacturing reached the highest seasonally adjusted level in two-and-a-half years, only 6% below the high recorded in July 2007.[1]

This cyclical recovery, however, has not stilled concerns about the sector's health. The number of U.S. manufacturing sites fell from 397,552 in 2001 to 344,352 as of June 2010, leaving many factories abandoned. Manufacturing employment, which peaked at 19.4 million in 1979, was 11.7 million in February 2011. Of those 11.7 million manufacturing workers, only 8.2 million, or 5.3% of the civilian labor force, are now engaged in factory production work.

These broad trends—generally expanding manufacturing output coupled with declining employment—are of long standing. In combination, they are taken as indicators of rapidly rising productivity. Labor productivity in U.S. manufacturing, defined as output per work hour, has increased 13% since 2005 and 41% since 2000 as manufacturers have shifted away from labor-intensive production. A rapid rise in productivity would be consistent with the belief that U.S. manufacturing is becoming more efficient and technologically sophisticated and therefore requires less labor; one analogy might be the farm sector, in which the labor force has shrunk to a small fraction of its size a century ago despite a vast increase in output.

The estimates of rising manufacturing output and capacity and of rapidly improving labor productivity, however, rely critically on price adjustments that attempt to account for improvements in the quality of computers and certain other high-technology products. Such adjustments are required because, for example, simply measuring changes in the quantity or value of the computers produced each year would have little economic meaning given the very rapid increase in a computer's capabilities.[2]

Government statistical agencies address this problem by making highly technical adjustments when measuring certain prices. These adjustments can affect prominent economic indicators, such as gross domestic product and labor productivity. The industries for which data are adjusted in this way, such as semiconductor manufacturing, are among the most vigorous in U.S. manufacturing, leading to questions about whether reported improvements in manufacturing represent real changes or merely the result of statistical adjustments.[3]

While some data thus indicate that U.S. manufacturing is resilient and recovering well from the 2007-2009 recession, two facts in particular support the argument that the manufacturing sector is more challenged than the government's output and productivity measures imply:

- Unlike previous expansions, the two most recent cyclical upturns in the U.S. economy have not brought more jobs in manufacturing. Factory output rose roughly 16% from the end of the 2001-2002 recession through 2007 without generating factory jobs. This pattern has repeated itself since June 2009; both total manufacturing employment and factory production employment are lower now than at the deepest point of the 2007-2009 recession, despite a 13.9% increase in factory output.
- Value added in U.S. factories appears to represent a declining share of the value of factory shipments. Although total output may still be rising, the measured contribution of U.S. factories to the value of the final products may not be keeping pace. Some commentators refer to this phenomenon as "hollowing out."

WHAT IS VALUE ADDED?

Value added represents one measure of the health of manufacturing. Conceptually, value added equals the value of manufacturers' shipments less the value of purchased inputs. Employees' pay and benefits, depreciation of capital investment, business income taxes, and returns to business owners all are components of value added. In essence, value added is meant to capture the share of the value of final products that is being added "in house." Value added is typically assessed with two different metrics. One metric, the growth rate of "real" value added, provides information about the expansion of industrial output but is subject to the technology-related adjustment issues discussed above. The other metric, which avoids these adjustment issues, is the ratio of each year's manufacturing value added to that year's manufacturing shipments.

For an individual firm, a decline in the ratio of value added to shipments may not be meaningful. To see why, consider a firm that produces a component, uses the component to make another product, and sells that product. If the firm were to split itself in half, so that one entity makes the component and sells it to a separate entity that makes the finished product, manufacturers' total shipments would increase but total value added would not change. The resulting decline in the ratio of value added to shipments would have no economic significance.

The situation may be different, however, at the level of an industry or of the manufacturing sector as a whole. In these cases, a lower ratio of value added to shipments could reveal important changes. One might be diminished profitability. Another might be that manufacturers' costs for certain inputs, such as electricity or paperboard cartons, are rising faster than the prices manufacturers receive for their products. A third possibility could be that manufacturers are collectively making greater use of imported parts and components.

From 1990 through 2005, U.S. manufacturing value added fluctuated in a narrow range, between 46.3% and 48.5% of the value of manufacturers' shipments, according to Census Bureau estimates. From 2006 through 2008, however, the ratio fell to unusually low levels (see Figure 1). The decline, which began in 2003, predates the 2007-2009 recession.

Data compiled by the Bureau of Economic Analysis (BEA), using different methods, show similar trends over time. Both data series show the manufacturing value-added ratio rising in 2009, but as the figures are subject to large revisions, it is premature to conclude that the value-added ratio has returned to its previous range.

Whether the declining ratio of value added to shipments indicates deterioration in U.S. manufacturing is a matter of controversy among economists, as a number of conceptual issues and statistical deficiencies make this an imperfect gauge of economic activity. Among the most important of these problems are the following.

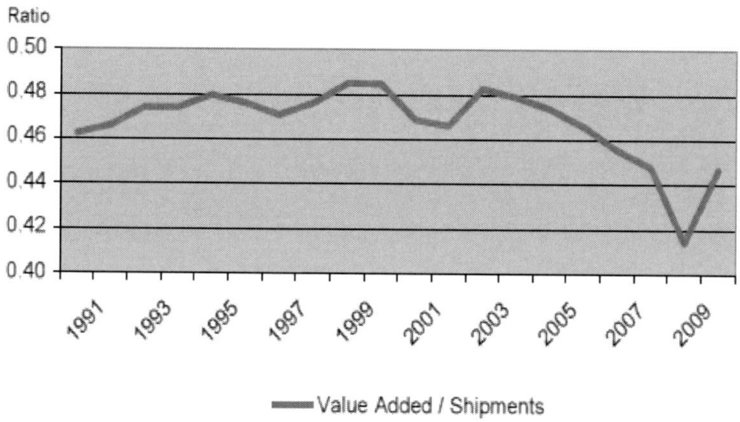

Source: U.S. Census Bureau, Annual Survey of Manufactures.

Figure 1. Manufacturers' Value Added; (Ratio to Shipments).

Accounting for Purchased Services

The Census Bureau *includes* manufacturers' outlays for telecommunications, advertising, transportation, and other services purchased from third parties in value added, but *excludes* outside purchases of materials, such as parts, packaging, and fuel. Conceptually, there is no reason to treat purchased services differently from purchased materials, but until recently the government lacked data on purchased services. According to BEA data, spending on purchased services consumed 16.2% of manufacturers' sales in 2006, up from 14.9% in 1997. This relative growth in the use of purchased services would have increased the reported growth rate of value added in manufacturing during that period, even if there were no actual change in physical manufacturing activity.[4]

Accounting for Research and Development

Government statistics include the costs of a manufacturer's research and development staff in value added in the same way as the costs of its production employees. Economists have long debated whether research and development should be treated instead as investment. BEA estimates that this accounting change would have raised the annual growth rate of value added in private industry slightly between 1995 and 2007, and that it would have caused value added in certain high-tech sectors, notably pharmaceutical, instrument, and aerospace manufacturing, to grow faster than official statistics indicate. BEA plans to revise the statistical treatment of corporate research and development spending in 2013.[5]

Intellectual Property Exports

Many companies conduct research and development in the United States and use or license the resulting designs, patents, and brand names for manufacturing abroad. In principle, if this intellectual property is licensed to a foreign producer, it appears in U.S. trade data as a services export. However, the measurement of value added can become blurred if the foreign-made product is then imported into the United States to be incorporated into other goods; U.S. data on manufacturing may not adequately correct for the fact that some of the import's value was originally created in the United States or may categorize that U.S. value added as a product of the service sector rather than the manufacturing sector. Moreover, if the intellectual property is licensed from a U.S. operation to a foreign operation of a multinational company, the licensing fee may not reflect the true economic value of the intellectual property. These complexities tend to make U.S. manufacturing value added appear smaller than it really is, and this bias may have increased over time as "offshoring" of assembly work has become more common.

Factoryless Manufacturing

A growing number of companies widely considered to be manufacturers—perhaps the best known is the electronics company Apple Inc.—specialize in certain processes, such as design, distribution, or service, but perform little or no physical production themselves.[6] The activities of such "factoryless manufacturers" may show up in government data as "wholesale trade" rather than as "manufacturing." As more companies embrace the factoryless model, it is possible that an increasing proportion of products with high U.S. value added are being omitted from the calculation of manufacturing output and value added, contributing to the measured decline in the value-added ratio through most of the past decade. The Census Bureau plans to address this issue by including factoryless manufacturers in the manufacturing sector starting in 2012, but this may cause other statistical distortions.[7]

Price Biases

Import price indexes play a critical role in measuring value added in manufacturing.

Around 40% of all imports are inputs for business use, such as parts and components, rather than consumer goods. Research suggests that import price indexes are slow to capture price changes when a manufacturer shifts from a domestic to a foreign supplier or from a supplier in one country to a supplier in another. These measurement problems may result in official data understating the use of imported components in U.S. factories, implying that value added in U.S. factories may be lower than statistics indicate.[8]

Table 1. Summary of Measurement Issues Affecting Value Added in Manufacturing

Measurement Issue	Effect on Measured Value Added
Inclusion of Purchased Services in Value Added	Positive
Treatment of In-House Research and Development as Expense Rather Than Investment	Negative
Misattribution of Intellectual Property Exports	Negative
Misclassification of Some Factoryless Manufacturing	Negative
Import Price Overstatement	Positive

Source: Interviews and literature reviewed by CRS.

Table 1 summarizes the effects of these various measurement issues on reported value added in the manufacturing sector. The net effect is ambiguous. Although the statistical problems are serious, it is uncertain, on balance, whether they collectively make value added larger or smaller, relative to manufacturers' shipments, and whether they change the growth rate of value added in manufacturing.

INTERNATIONAL TRADE AND VALUE ADDED

One possible reason for the decline in manufacturing value added, relative to sales, could be increased use of imported inputs—so-called "intermediate inputs"—by U.S. manufacturers. In 1998, 24% of intermediate inputs used in manufacturing were imported. According to one analysis, the figure started rising in 2003 and reached 34% in 2006. Moreover, U.S. factories' use of domestic components and other materials (excluding energy) is estimated to have declined at an annual rate of 3.9% between 1998 and 2006, while their use of imported components and materials is estimated to have risen at a 3.5% rate.[9]

The rapidly increasing use of imported inputs may inflate labor productivity growth in manufacturing, as measured by government data. The reason for this is that government statistics may understate the declines in input prices if manufacturers are shifting quickly from using a domestic input to a competing foreign-made input that is lower in cost. If this is occurring, it would mean that U.S. factories are using a greater quantity of the foreign input than assumed, and less of the domestically made alternative. In that case, the output of U.S. manufacturing workers, and hence their productivity, would be lower than believed.[10]

Industry-level data suggest that increased use of imported components may be occurring in some manufacturing industries that traditionally have high value added relative to shipments. These industries typically are intensive users of scientific research and advanced technology, and are often regarded as industries in which the United States should have an international competitive advantage. While some such industries have shown little change in the domestic value added ratio, others have exhibited marked declines in value added relative to shipments in the last four or five years (see Table 2). For example, value added in the computer and electronics industry, with shipments of $328 billion in 2009, declined from 65% of shipments in 2005 to 59% in 2009.

Table 2. Value Added in Five High-Value-Added Industries (Percent of Value of Industry's Shipments)

Industry	2009 Shipments ($bn)	2000	2005	2008	2009 (prelim.)
Pharmaceuticals	$142	70.3%	82.4%	74.6%	74.8%
Medical Instruments	$35	74.8%	77.0%	72.4%	72.0%
Semiconductors	$59	74.7%	80.7%	74.6%	69.9%
Navigational & Control Instruments	$125	62.3%	71.0%	61.5%	63.5%
Dies and Tools	$6	69.3%	69.7%	62.1%	60.9%

Source: U.S. Census Bureau, Annual Survey of Manufactures.

The declining share of domestic value added in particular industries is related to a broad change in businesses' strategies that emphasizes the use of global supply chains.[11] In such arrangements, made possible by low freight transportation and communication costs, a retailer or manufacturer organizes its production on a worldwide basis rather than on a country-by-country basis. It may then obtain economies of scale in manufacturing by using a factory in one country to supply most or all of its need for a particular product worldwide, shipping intermediate inputs from place to place for additional processing in order to deliver the final product at the lowest total cost.

Global supply chains have enabled U.S. manufacturers to make greater use of imported intermediate inputs from low-wage countries, notably China. The trade patterns can be difficult to trace, because it may not be clear which domestic manufacturing industry makes use of a particular import. One recent study illustrates the complexities. Imports of automobile seats have declined since 1994, suggesting, at first glance, that auto manufacturers are making greater use of U.S. content. More detailed analysis, however, shows that imports of seat parts, mainly from Mexico, have increased sharply. These imports consist of items such as fabrics and temperature-control devices that may not be readily identified as auto-related.[12] Collectively, such imports reduce the amount of U.S. content in the seat and in the vehicle in which the seat is installed.[13]

The extent to which inputs imported into the United States contain value added in the exporting country, the United States, or third countries is uncertain due to the same conceptual factors that complicate analysis of U.S. value added in manufacturing. Nor are there reliable data on how the amount of U.S. content embedded in imported products has changed over time. However, one recent study estimated that 8.3% of the gross value of U.S. imports in 2004 originated in the United States. Among the world's major economies, only the European Union approached the United States in terms of domestic value embedded in imports. Developing economies usually add little value to exports that are subsequently incorporated into their imports (see Table 3).[14]

Table 3. Domestic Value Added Included in Imported Goods, 2004 (Percent of Each Country's Imports by Value)

United States	8.3%
European Union	7.2%
Japan	3.4%
South Korea	1.1%
China	0.9%
Brazil	0.4%

Source: Koopman et al., p. 43. See note 14.

One evidence of the tenuous link between output and value added can be seen in China's soaring exports of what U.S. trade data label "advanced technology products," or ATP, including specified electronic and biotechnology goods. While China's bilateral trade surplus with the United States in such products soared from 2002 through 2006, all of the increase was due to processing of foreign components in Chinese factories owned, at least in part, by foreign investors. Although U.S. exports to China were lower than Chinese exports to the United States, "It appears that ATP exports from the United States to China are dominated by large scale, sophisticated, high-valued equipment and devices at the high end of these industries' value-added chains, while ATP exports from China to the United States are mainly small scale final products or components in the low end of the ATP value-added chain," a recent study concluded.[15]

More generally, the so-called emerging economies tend to rely more heavily on foreign value added for export goods than do higher-income countries (see Table 4). The sources of that imported value added vary considerably among countries. The United States, for example, is estimated to have provided 37.1% of the value added in Mexican exports in 2004, whereas Japan provided only 2.7%. On the other hand, Japan was the largest source of foreign value in China's exports, providing 7.9% of total value added, while the United States provided only 3.7%.[16]

Table 4. Domestic Value Added in Final Goods Exports (Percent of Export Value, 2004)

European Union (15 countries)	88.9%
Japan	88.8%
United States	88.4%
Korea	69.3%
China	64.1%
Mexico	56.5%

Source: Koopman et al., p. 45. See note 14.

These data shed light on a concern frequently raised in the context of national security, that U.S. manufacturers of vital products are critically dependent upon inputs from abroad. Evidence suggests that the output of U.S. factories contains a large proportion of foreign value added. However, many other countries are even more dependent upon foreign value added than is the United States, at least with respect to goods traded in international markets.

ISSUES FOR CONGRESS

As the research surveyed in this report emphasizes, traditional understandings of "manufacturing" are inadequate to explain the process by which goods are produced in the modern world economy. For a large number of goods, physical production—activities such as stamping, molding, cutting, machining, welding, and assembly—is no longer the heart of the manufacturing process. The bulk of the value in many goods comes from ancillary activities, such as design, marketing, and distribution, which are not necessarily performed by the same enterprises, or at the same locations, as physical production.

The shift to global supply chains has had both positive and negative effects on the U.S. economy. There is no doubt that it has contributed to reduced U.S. consumer prices for many manufactured products. The availability of imported intermediate inputs has probably preserved some manufacturing within the United States, as reliance on higher-cost domestic inputs might well make related U.S. final-goods manufacturing uncompetitive. Additionally, the supply chains themselves support U.S. jobs in transportation, logistics management, and other fields.

At the same time, there is widespread agreement that "offshoring" has played a major role in loss of factory production work (see Figure 2), leading to higher unemployment and reduced incomes for some groups of workers and some communities where import-sensitive manufacturing is located.[17] One forthcoming study estimates that the rapid growth of manufactured imports from China accounted for a quarter or more of the decline in U.S. manufacturing employment since 1990, and may have adversely affected employment and wages in other sectors as well.[18]

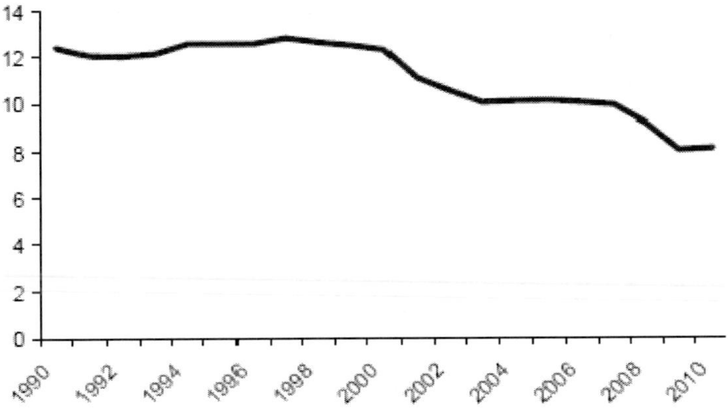

Source: Bureau of Labor Statistics, Current Employment Statistics.
Note: Data are for December of each year, seasonally adjusted.

Figure 2. U.S. Production Employment in Manufacturing; (Million Workers).

The broader impact on the U.S. labor market, however, remains a matter of considerable debate. One recent study of the growth of Chinese exports to the European Union between 1999 and 2007, directly applicable to the United States, concludes that "trade drives out low-tech firms ... and increases the incentives of incumbents to speed up technical change."[19] This finding, if confirmed in other research, may require reevaluation of the many economic studies that attribute declining factory employment to technological change, as it emphasizes that the rate of technological change is itself affected by increased trade with low-wage countries. It also suggests that declines in manufacturing production employment may go hand in hand with increased demand for workers with skills that are in some way related to goods production and distribution, but may not fall within the traditional definition of "manufacturing" work.

The transformation of manufacturing poses novel issues for public policies aimed at the manufacturing sector. A variety of federal programs, from the Hollings Manufacturing Extension Partnership administered by the National Institute of Standards and Technology[20] to the National Nanotechnology Initiative[21] to the Small Business Administration's 504/CDC Loan Guaranty Program,[22] are designed, in part, to help manufacturers upgrade technology, replace capital stock, and compete more effectively in global markets. The extent to which such efforts lead private-sector firms to select U.S. locations for high-value activities within their supply chains, and the degree to which those activities create employment, are unclear.

More broadly, shifts in the nature of value added in manufacturing put into question the efficacy of policies designed to promote factory production within the United States, such as tax policies favoring investment in manufacturing equipment and "Buy American" rules requiring certain goods financed by the federal government to be produced domestically. Given that employment and economic growth are increasingly decoupled from production, it is uncertain whether policies oriented to physical manufacturing activity are best suited to achieve desired economic goals.

Finally, the changes described in this report raise questions about the adequacy of government statistics. U.S. statistical agencies have made significant efforts in recent years to improve the collection of data on the service sector, corporate spending on research and development, and international trade in intangible products. Nonetheless, available data still tend to treat manufacturing and services as unrelated economic activities, and it is not clear that existing data series on domestic economic activity, trade, and freight transportation completely capture changes in the nature of manufacturing, the sources of employment, and the creation of value.

End Notes

[1] Industrial production and capacity data are published in the Federal Reserve Board's monthly G.17 release, available at http://www.federalreserve.gov/releases/G17/Current/default.htm.

[2] For an explanation of this adjustment process, see Dave Wasshausen and Brent R. Moulton, "The Role of Hedonic Methods in Measuring Real GDP in the United States," October 12-13, 2006, http://www.bea.gov/papers/pdf/ hedonicGDP.pdf.

[3] David Byrne, Brian K. Kovak, and Bryan Michaels, "Offshoring and Price Measurement in the Semiconductor Industry," in Susan N. Houseman and Kenneth F. Ryder, eds., *Measurement Issues Arising from the Growth of Globalization* (Washington, DC: National Academy of Public Administration, 2010), pp. 169-194.

[4] Robert E. Yuskavage, Eric H. Strassner, and Gabriel W. Medeiros, "Domestic Outsourcing and Imported Inputs in the U.S. Economy: Insights from Integrated Economic Accounts," May 15, 2008, http://www.bea.gov/papers/pdf/ yuskavage_outsource.pdf, p. 40.

[5] Jennifer Lee and Andrew G. Schmidt, "Research and Development Satellite Account Update," *Survey of Current Business*, December 2010, pp. 16-27.

[6] Apple, Inc., Form 10-K for the fiscal year ended September 25, 2010, p. 13, states that "substantially all of the Company's components and products are manufactured in whole or in part by a few third-party manufacturers." The only company-owned factory mentioned is in Cork, Ireland.

[7] U.S. Census Bureau, "Economic Classification Policy Committee (ECPC) Recommendation for Classification of Outsourcing in North American Industry Classification System (NAICS) Revisions for 2012," May 12, 2010, http://www.census.gov/eos/www/naics/fr2010/ECPC%20Recommendation%20for%20Classification%20of%20Outsourcing.doc.

[8] Susan N. Houseman, "Offshoring and Import Price Measurement," *Survey of Current Business*, February 2011, pp. 7-11.

[9] Lucy P. Eldridge and Michael J. Harper, "Effects of imported intermediate inputs on productivity," *Monthly Labor Review*, June 2006, pp. 6, 12.

[10] Susan Houseman, Christopher Kurz, Paul Lengerman, and Benjamin Mandel, "Offshoring Bias in U.S.
Manufacturing," *Journal of Economic Perspectives*, forthcoming.

[11] CRS Report R40167, *Globalized Supply Chains and U.S. Policy*, by Dick K. Nanto. A growing academic literature examines global supply chains. Among the early examples is Gary Gereffi and Miguel Korzeniewicz, eds., *Commodity Chains and Global Capitalism* (Westport, CT: Greenwood Press, 1994).

[12] Thomas H. Klier and James R. Rubenstein, "Imports of Intermediate Parts in the Auto Industry—A Case Study," in Houseman and Ryder, eds., *Measurement Issues Arising from the Growth of Globalization*, pp. 226, 231.

[13] Value added in the motor vehicle parts sector, which averaged 41.4% of shipments from 1997 through 2006, averaged only 37.2% of shipments from 2007 through 2009, according to Census Bureau data.

[14] Robert Koopman, William Powers, Zhi Wang, and Shang-Jin Wei, "Giving Credit Where Credit Is Due: Tracing Value Added in Global Production Chains," Working Paper 16426, National Bureau of Economic Research, September 2010, revised March 2011, http://www.nber.org/papers/w16426.pdf.

[15] Michael Ferrantino, Robert Koopman, Zhi Wang, Falan Yinug, Ling Chen, Fengjie Qu, and Haifeng Wang, "Classification and Statistical Reconciliation of Trade in Advanced Technology Products," May 2008, http://ssrn.com/ abstract=1132748, pp. 38-43, 51. For a study reaching similar conclusions with respect to the sophistication of Chinese exports, see Mary Amiti and Caroline Freund, "The Anatomy of China's Export Growth," World Bank Development Research Group, Policy Research Working Paper 4628, May 2008,

http://www-wds.worldbank.org/external/default/main?pagePK=64193027&piPK=64187937&theSitePK=523679&menuPK=64187510&searchMenuPK=64187283&theSitePK=523679&entityID=000158349_20080527092730&searchMenuPK=64187283&theSitePK=523679.

[16] Koopman et al., "Giving Credit Where Credit Is Due," p. 45.

[17] For a summary of relevant research on employment and "offshoring," see CRS Report RL32292, *Offshoring (or Offshore Outsourcing) and Job Loss Among U.S. Workers*, by Linda Levine.

[18] David Autor, David Dorn, and Gordon H. Hanson, "The China Syndrome: Local Labor Market Effects of Import Competition in the U.S.," National Bureau of Economic Research, Working Paper, March 2011.

[19] Nicholas Bloom, Mirko Draca, and John Van Reenen, "Trade Induced Technical Change? The Impact of Chinese Imports on Innovation, IT and Productivity," National Bureau of Economic Research, Working Paper 16717, January 2011, p. 4.

[20] CRS Report 97-104, *Manufacturing Extension Partnership Program: An Overview*, by Wendy H. Schacht.

[21] CRS Report RL34401, *The National Nanotechnology Initiative: Overview, Reauthorization, and Appropriations Issues*, by John F. Sargent Jr.

[22] CRS Report R41184, *Small Business Administration 504/CDC Loan Guaranty Program*, by Robert Jay Dilger.

In: United States Manufacturing
Editors: V. Murphy and J. E. Walker
ISBN: 978-1-62081-522-9

Chapter 3

THE IMPACT OF EXPORTING ON THE STABILITY OF UNITED STATES MANUFACTURING INDUSTRIES*

David Riker and Brandon Thurner

ACKNOWLEDGMENTS

We appreciate the helpful comments that we received from Jeffrey Anspacher, Joseph Flynn, Martin Johnson, Don Lee, Michael Nicholson, and William Powers. All remaining errors are our own.

EXECUTIVE SUMMARY

Exporting provides opportunities to expand the sales of U.S. manufacturing industries, resulting in greater revenues for U.S. companies and higher employment for U.S. workers. However, investors and workers are concerned about risk as well as expected return, and risk depends on the volatility of economic outcomes. In the case of exports, it is worth asking whether the added risks and unpredictability associated with exporting outweigh the increase in expected revenues.

---

* This is an edited, reformatted and augmented version of International Trade AQdministration, Manufacturing and Services Economics Brief, dated March 2011.

In theory, exporting could increase the volatility of the total shipments of U.S. manufacturers by increasing cost competition and by exposing the manufacturers to other types of uncertainty, including exchange rate volatility and fluctuations in shipping costs. On the other hand, exporting could reduce the volatility of their total shipments by diversifying shocks to product demand.

In this paper, we investigate the recent link between exporting and the economic stability of the U.S. manufacturing sector. We compare the volatility of the domestic shipments of U.S. companies and industries to the volatility of their total, worldwide shipments.

First, we analyze the 2000-2009 revenues reported in the public financial statements of two U.S. companies with substantial non-U.S. sales, the Xerox Corporation and the Dow Chemical Company. Both case studies indicate that the companies' non-U.S. revenues reduce the volatility of their total, worldwide revenues.

Second, we analyze industry-level shipments data for the U.S. manufacturing sector from the Annual Survey of Manufactures and the Economic Census for the period 2000-2008. The industry-level data indicate that exporting reduces the volatility of total shipments for a significant majority of the U.S. manufacturing industries. In general, industries with higher export shares experienced larger reductions in the volatility of their total shipments.

1. Introduction

Exporting provides opportunities to expand the sales of U.S. manufacturing industries, resulting in greater revenues for U.S. companies and higher employment for U.S. workers. However, investors and workers are concerned about risk as well as expected return, and risk depends on the volatility of economic outcomes. In the case of exports, it is worth asking whether the added risks and unpredictability associated with exporting outweigh the increase in expected revenues.

In theory, exporting could increase the volatility of the total shipments of U.S. manufacturers by increasing cost competition and by exposing the manufacturers to other types of uncertainty, including exchange rate volatility and fluctuations in shipping costs. On the other hand, exporting could reduce the volatility of the total shipments of U.S. manufacturers by diversifying shocks to product demand. If the second effect dominates, and the volatility of the total shipments of U.S. manufacturers are less than the volatility of their U.S. shipments, then exporting reduces the volatility of the manufacturers' revenue streams. Reducing the volatility of revenues has clear economic benefits. It is valuable to companies and workers that prefer income stability. In addition, reduced volatility can encourage investment in new technologies, physical capital, and human capital.[1]

In this paper, we investigate the recent empirical link between exporting and the economic stability of U.S. manufacturing industries. To measure the volatility of the shipments of U.S. manufacturers, we calculate the ratio of the standard deviation of the unpredictable component of shipments to the most recent value of shipments. We call this ratio the *normalized volatility* of shipments. Since the standard deviation and the most recent value of shipments are measured in the same units, the units cancel from the ratio. This conversion to a percentage, or normalization, facilitates comparison of the volatility measures among companies and industries that vary significantly in size.

We compare the volatility of the domestic shipments of U.S. companies and industries to the volatility of their total, worldwide shipments. First, we examine the 2000-2009 revenues reported in the public financial statements of two U.S. companies with substantial non-U.S. sales, the Xerox Corporation and the Dow Chemical Company.[2] Then we build on these company case studies by analyzing industry-level shipments data for the U.S. manufacturing sector from the Annual Survey of Manufactures (ASM) and the Economic Census for the period 2000-2008.

2. COMPANY CASE STUDY: THE XEROX CORPORATION

The Xerox Corporation (Xerox) is a global document management company that was founded in 1906 and is currently headquartered in Norwalk, Connecticut. [3] The company employs approximately 130,000 people worldwide. It serves non-U.S. markets through a combination of U.S. exports and foreign production, including joint ventures. The company's product offerings include copiers, fax machines, and printers, as well as related services and supplies. The company identifies the reputation of its brands and its innovative technologies as the keys to its international success. Xerox publicly reports its annual revenues for three geographic areas: the United States, Europe, and Other Areas. It is important to note that the company's non-U.S. revenues are not exclusively exports from the United States. They include sales of goods and services produced by the company's foreign affiliates.[4] The company's exports from the United States are not separately quantified in the company's Annual Reports. This is a limitation on the usefulness of publicly available company data for analyzing the link between exports and the volatility of revenues. (However, this limitation does not apply to the industry-level analysis that we report below.) Table 1 reports the company's 2000-2009 revenues by geographic area. In 2009, U.S. revenues accounted for 53.7% of the total revenues of Xerox.

Table 1. Annual Revenues of the Xerox Corporation by Geographic Segment

Year	U.S. revenues in millions of constant 2008 dollars	Total revenues in millions of constant 2008 dollars	U.S. share of the company's total revenues
2000	13,102	22,947	57.10%
2001	12,008	20,354	59.00%
2002	10,712	18,665	57.39%
2003	9,853	18,100	54.44%
2004	9,356	17,625	53.08%
2005	9,100	17,033	53.43%
2006	8,831	16,699	52.88%
2007	9,272	17,596	52.69%
2008	9,122	17,608	51.81%
2009	8,060	15,000	53.73%

Source: Annual Reports of the Xerox Corporation.

Between 2000 and 2009, the company's annual U.S. revenues ranged from $8.06 billion to $13.10 billion in constant 2008 dollars.[5] Its annual total revenues ranged from $15.00 billion to $22.95 billion over this period. The normalized volatility of the company's total revenues (7.52%) was smaller than the normalized volatility of its U.S. revenues (8.53%), so the company's non-U.S. revenues reduced the normalized volatility of its total revenues.[6]

3. Company Case Study: The Dow Chemical Company

The Dow Chemical Company (Dow) is one of the largest chemical producers in the world.[7] It was founded in 1897 and is headquartered in

Midland, Michigan. Dow manufactures and sells chemicals, plastic materials, agricultural products and services, advanced materials and other specialized products and services.

The company employs approximately 52,000 people worldwide. The company's success in technological innovation is fueled by annual R&D expenditures in excess of $1 billion. Dow identifies the geographic balance of its sales as a key to its success. Dow reports its annual revenues for three geographic areas: the United States, Europe, and the Rest of World. Again, it is important to note that the company's non-U.S. revenues are not exclusively exports from the United States. Table 2 reports the company's 2000-2009 revenues by geographic area.

Table 2. Annual Revenues of the Dow Chemical Company by Geographic Segment

Year	U.S. revenues in millions of constant 2008 dollars	Total revenues in millions of constant 2008 dollars	U.S. share of the company's total revenues
2000	15,971	36,466	43.80%
2001	14,355	33,598	42.72%
2002	13,259	32,514	40.78%
2003	14,771	37,620	39.27%
2004	16,876	45,022	37.48%
2005	19,011	50,235	37.84%
2006	19,092	51,610	36.99%
2007	18,520	54,515	33.97%
2008	18,306	57,361	31.91%
2009	13,978	44,346	31.52%

Source: Annual Reports of the Dow Chemical Company.

In 2009, U.S. revenues accounted for 31.5% of the total revenues of Dow. Between 2000 and 2009, the company's annual U.S. revenues ranged from $13.26 billion to $19.09 billion in constant 2008 dollars. Its annual total revenues ranged from $32.51 billion to $57.36 billion.

The normalized volatility of the company's total revenues (11.93%) was smaller than the normalized volatility of its U.S. revenues (14.55%), so the company's non-U.S. revenues reduced the volatility of its total revenues.

4. Analysis of 85 Manufacturing Industries in the U.S. Manufacturing Sector

While the two company case studies provide a measure of the reduction in the normalized volatility of total revenues that results from adding non-U.S. sales to U.S. sales, they do not specifically quantify the contribution of exports from the United States.

However, using industry-level data, we can focus the comparisons of normalized volatilities on the shipments of *U.S. production* to the United States and foreign markets. The industry-level analysis also has the advantage that it covers the entire U.S. manufacturing sector, and so it significantly expands the breadth of the analysis.

We analyze publicly available four-digit NAICS industry data on the value of shipments from the ASM for 2000-2001, 2003-2006, and 2008 and from the Economic Census for 2002 and 2007. At this level of disaggregation, there are 85 industries in the U.S. manufacturing sector.

The ASM is based on a subset of the population of establishments in the Economic Census. It includes approximately 55,000 manufacturing establishments, with 10,000 large establishments selected with certainty and another 45,000 other establishments selected with probability proportional to establishment size. To be included in the ASM, a manufacturing establishment must have one or more paid employees or leased employees engaged in activities that are classified in NAICS industries 31 through 33.

In order to calculate the domestic shipments of the U.S. industries, we supplement the ASM data with statistics on the free alongside ship value of U.S. domestic exports from the U.S. ITC's Trade Dataweb.[8] We calculate the U.S. shipments of each industry in each year as the difference between its total value of shipments and its domestic exports. Across the 85 industries, the ratio of exports to total shipments in 2008 ranged from 0.6% to 71.7%, with a simple average of 19.1%.

Using these data, we estimate the contribution of exporting to the normalized volatility of the total value of shipments of the U.S. manufacturing industries. Before we report the volatility comparisons for all 85 industries, we will go through a detailed example for a single industry, the Medical Equipment and Supplies Manufacturing industry (NAICS 3391).

In 2008, this industry had U.S. shipments of $64.09 billion, and total shipments of $84.03 billion (Table 3).

Table 3. Example of NAICS Industry 3391 (Medical Equipment and Supplies) Revenues in Billions of 2008 U.S. Dollars

Year	U.S. Shipments Only	Total Shipments	U.S. Share
2000	54.36	66.00	82.36%
2001	56.51	69.13	81.74%
2002	59.86	72.74	82.29%
2003	61.45	75.69	81.19%
2004	59.60	74.86	79.62%
2005	63.77	80.52	79.20%
2006	65.00	82.46	78.83%
2007	61.45	79.81	77.00%
2008	64.04	84.03	76.21%
Ratio of Standard Deviation to 2008 Value	0.029	0.021	

Source: Annual Survey of Manufactures 2000-2001, 2003-2006, 2008; Economic Census 2002, 2007; ITC Trade Dataweb.

The difference between these two numbers is the industry's exports. Between 2000 and 2008, the standard deviation of unpredictable component of the industry's U.S. shipments was $1.78 billion, and the standard deviation of the unpredictable component of its total shipments was $1.84 billion. Although the standard deviation was greater for the U.S. industry's total shipments than for its U.S. shipments, the percentage increase in the standard deviation from expanding the market (3.37%, or 1.84/1.78-1) was smaller than the percentage increase in the 2008 value of shipments (43.12%, or 69.17/48.33-1). It follows that the normalized volatility of the industry's total shipments (2.1%) is less than the normalized volatility of the industry's U.S. shipments (2.9%). We conclude from this comparison that exporting reduced the volatility of the industry's total shipments.[9]

The normalized volatility comparisons for the Medical Equipment and Supplies industry are representative of the results for the 85 manufacturing industries as a whole (The results for each of the industries are reported in a table in the Technical Appendix.). For 21 of the 85 industries, both the normalized volatility and the standard deviation of the industries' total shipments were smaller than their counterparts for the industries' U.S. shipments. The smaller standard deviations for these industries indicate that exports were a particularly good hedge against fluctuations in demand in the U.S. market.[10] For an additional 40 of the 85, the normalized volatility measure was lower for total shipments than for U.S. shipments (but the standard deviation was not), which indicates that exporting reduced the volatility of the shipments of these additional industries relative to their 2008 values.

Among the 85 industries, the Leather & Hide Tanning & Finishing, Audio & Video Equipment, Apparel Accessories & Other Apparel, Communications Equipments and Engine & Turbine & Power Transmission Equipment industries experienced the largest percentage reductions in the normalized volatility of their total shipments as a result of their exports.[11] In general, industries with higher export shares experienced larger reductions in the normalized volatility of their total shipments.

As a sensitivity analysis, we recalculated the normalized volatilities using the mean value of shipments between 2000 and 2008 in the denominator rather than the 2008 value of shipments. The results are similar. For 21 of the 85 industries, both the normalized volatility and the standard deviation of the industries' total shipments were smaller than their counterparts for the industries' U.S. shipments. For an additional 33 of the 85, the normalized volatility measure was lower for total shipments than for U.S. shipments (but the standard deviation was not), which indicates that exporting reduced the volatility of the shipments of these additional industries relative to their 2008 values.

Finally, we also examined whether the results are different when we focus the analysis on exports to OECD member countries. We defined a U.S. industry's OECD shipments as the sum of its U.S. shipments and its exports to other OECD countries. Of the 85 industries, there are 55 industries for which exports to all countries reduced the normalized volatility of their shipments and exports to other OECD countries also reduced the normalized volatility. There are six industries for which total exports reduced the normalized volatility, but exports to other OECD countries did not. There are two industries for which exports to other OECD countries reduced the normalized volatility of their shipments, but total exports did not.

CONCLUSION

Both the company case studies and the industry-level analysis of the U.S. manufacturing sector indicate that non-U.S. sales, including exports from the United States, reduce the normalized volatility of the value of total shipments of U.S. manufacturers. The value of shipments is a measure of U.S.

manufacturers' revenues and not their profits, so the reduction in the volatility of shipments does not necessarily imply a reduction in the volatility of the companies' profits. However, the two are closely connected as long as costs are a steady share of revenues.

REFERENCES

Bacchetta, Philippe and Eric van Wincoop (2000): "Does Exchange Rate Stability Increase Trade and Welfare?" *American Economic Review* 90(5): 1093-1109.

Bahmani-Oskooee, Mohsen and Scott W. Hegerty (2009): "The Effects of Exchange-Rate Volatility on Commodity Trade between the United States and Mexico." *Southern Economic Journal* 75(4): 1019-1044.

Dixit, Avinash K. and Robert S. Pindyck (1994): *Investment under Uncertainty*. Princeton, N.J.: Princeton University Press.

Ekanayake, E.M., John R. Ledgerwood, and Sabrina D'Souza (2010): "The Real Exchange Rate Volatility and U.S. Exports: An Empirical Investigation." *International Journal of Business and Finance Research* 4(1): 23-35.

Helpman, Elhanan and Paul Krugman (1985): *Market Structure and Foreign Trade*. Cambridge, Massachusetts: MIT Press.

Koray, Faik and William D. Lastrapes (2001): "Real Exchange Rate Volatility and U.S. Bilateral Trade: A VAR Approach." *Review of Economics and Statistics*, 708-712.

TECHNICAL APPENDIX

This appendix provides further details on some of the more technical aspects of the analysis in the paper.

Adjustments to the Shipments Data to Remove Predictable Trends

In this analysis, we are measuring the unpredictable variation in the shipments series. We are not trying to measure all time series variation in the shipments series, which would include variation in shipments over time due to a deterministic linear trend in the series. For each of the series, we estimated industry-specific means and linear trends for the 2000-2008 sample using Ordinary Least Squares. These simple regression models indicate that there is a positive trend in each series that is statistically significant at the 1% level. On this basis, we conclude that the expected values of the variables are not fixed over time. We calculated the unpredictable component of each of the shipments series as the residuals of its linear trend model.

Next we calculate the standard deviation, which is the square root of the variance. The general definition of variance is that it is the average squared difference between a variable and its expected value. If the expected value happens to be constant over time, then this is the average squared difference between the variable and its time-invariant mean. For these shipments series, however, the expected value is not time-invariant. It increases over time according to the linear trend models. In each year, the actual value of the variable differs from the expected value due to unpredictable variation. When we de-trend the shipment series by subtracting the expected value from the actual value, we are isolating the unpredictable variation, which has mean zero by construction over the 2000-2008 sample period. The de-trending does not require an economic theory about the determinants of the trends or assumptions about whether the trends are the same for U.S. shipments and total shipments or across industries. We measure the normalized volatility of the shipments data as the ratio of the standard deviation of the residuals to the most recent value of the shipments variable, which is the value in 2008. As a sensitivity analysis, we re-calculate the ratio using the average value for shipments between 2000 and 2008 as the denominator. Our measure of normalized volatility is similar to the conventional coefficient of variation statistic. However, the coefficient of variation is based on the standard deviation of the original series, not the standard deviation of the unpredictable components of the series. We modify the coefficient of variation statistic to address the predictable trends in the shipment series.

Economic Interpretation of the Comparison of Normalized Volatilities

For industries in which the normalized volatility of total shipments was lower than the normalized volatility of U.S. shipments, we infer that exporting reduces the volatility of the industry's total shipments. This interpretation is based on a counterfactual comparison with specific assumptions. We are assuming that the volatility of the industry's U.S. shipments would be the same in the absence of exports (a hypothetical scenario) as it is with exports (the actual, observed outcome).

This is implied by a more technical assumption – that the U.S. industry's marginal cost of production is separable between shipments to the different geographic markets. As long as prices are set based on the U.S. manufacturers' marginal costs, exporting will only affect the value of shipments to the U.S. market if it affects the marginal cost of supplying the U.S. market. If marginal costs are separable between the destination markets, then exporting will not affect the value of shipments to the U.S. market, even if there are fixed costs of production that are shared across the destination markets or fixed costs of accessing export markets.

This technical assumption is consistent with the most common economic models of international trade. It holds if production costs exhibit constant returns to scale, as in the classical Heckscher-Ohlin and Ricardian models of international trade, and the industry is not large enough to have a significant impact on factor prices.

It also holds for models with increasing returns to scale and imperfect competition, as long as there are constant marginal costs, as is the case in Helpman and Krugman (1985) and many related models of international trade.

Relation to the Economics Literature

Most of the economics literature that relates volatility to international trade focuses on a single source of uncertainty, exchange rate volatility. Examples of this literature include Bacchetta and van Wincoop (2000), Koray and Lastrapes (2001), Bahmani-Oskooee and Hegerty (2009), and Ekanayake et al. (2010).

In contrast, the comparison of normalized volatilities in this paper avoids disentangling the contributions of different sources of uncertainty that are associated with exporting, like fluctuations in aggregate demand in the foreign country, shipping costs, and exchange rates. We demonstrate in this paper that it is not necessary to quantify the separate contribution of each source of volatility in order to determine whether exporting reduced the volatility of an industry's total shipments.

Appendix Table: Ratio of the Standard Deviation to the Value in 2008

NAICS	Industry	U.S. Shipments Only	Total Shipments	U.S. Share
3111	Animal food mfg	0.061	0.059	95.34%
3112	Grain & oilseed milling	0.080	0.080	87.55%
3113	Sugar & confectionery product mfg	0.038	0.033	93.23%
3114	Fruit & vegetable preserving & specialty food mfg	0.024	0.025	92.81%
3115	Dairy product mfg	0.040	0.042	96.01%
3116	Animal slaughtering & processing	0.024	0.020	90.34%
3117	Seafood product preparation & packaging	0.043	0.041	95.95%
3118	Bakeries & tortilla mfg	0.015	0.015	97.61%
3119	Other food mfg	0.023	0.020	92.45%
3121	Beverage mfg	0.023	0.020	95.50%
3122	Tobacco mfg	0.114	0.117	97.80%
3131	Fiber, yarn, & thread mills	0.076	0.061	83.05%
3132	Fabric mills	0.129	0.075	63.08%
3133	Textile & fabric finishing & fabric coating mills	0.084	0.069	87.08%
3141	Textile furnishings mills	0.128	0.113	90.39%
3149	Other textile product mills	0.077	0.076	89.99%
3151	Apparel knitting mills	0.209	0.189	87.59%
3152	Cut & sew apparel mfg	0.235	0.234	87.23%
3159	Apparel accessories & other apparel mfg	0.512	0.273	42.59%
3161	Leather & hide tanning & finishing	1.002	0.259	28.30%
3162	Footwear mfg	0.295	0.244	75.53%
3169	Other leather & allied product mfg	0.233	0.201	61.78%
3211	Sawmills & wood preservation	0.137	0.127	90.44%
3212	Veneer, plywood, & engineered wood product mfg	0.177	0.164	92.57%
3219	Other wood product mfg	0.089	0.086	97.02%
3221	Pulp, paper, & paperboard mills	0.039	0.049	84.42%
3222	Converted paper product mfg	0.035	0.032	90.86%
3231	Printing & related support activities	0.042	0.042	93.47%
3241	Petroleum & coal products mfg	0.082	0.084	92.43%
3251	Basic chemical mfg	0.065	0.061	77.11%
3252	Resin, syn rubber, & artificial syn fibers & filaments mfg	0.094	0.075	65.85%
3253	Pesticide, fertilizer, & other agricultural chemical mfg	0.046	0.060	74.12%
3254	Pharmaceutical & medicine mfg	0.034	0.022	78.54%
3255	Paint, coating, & adhesive mfg	0.031	0.030	89.03%
3256	Soap, cleaning compound, & toilet preparation mfg	0.039	0.036	87.86%
3259	Other chemical product & preparation mfg	0.053	0.054	79.92%
3261	Plastics product mfg	0.036	0.035	90.29%
3262	Rubber product mfg	0.041	0.040	79.83%

NAICS	Industry	U.S. Shipments Only	Total Shipments	U.S. Share
3271	Clay product & refractory mfg	0.061	0.052	79.55%
3272	Glass & glass product mfg	0.027	0.028	80.49%
3273	Cement & concrete product mfg	0.075	0.075	99.41%
3274	Lime & gypsum product mfg	0.177	0.169	96.19%
3279	Other nonmetallic mineral product mfg	0.045	0.043	88.51%
3311	Iron & steel mills & ferroalloy mfg	0.081	0.079	86.64%
3312	Steel product mfg from purchased steel	0.087	0.086	98.22%
3313	Alumina & aluminum production & processing	0.081	0.084	83.24%
3314	Nonferrous metal (except aluminum) production & processing	0.116	0.115	49.36%
3315	Foundries	0.067	0.068	96.79%
3321	Forging & stamping	0.094	0.093	98.72%
3322	Cutlery & handtool mfg	0.072	0.063	76.93%
3323	Architectural & structural metals mfg	0.063	0.063	97.56%
3324	Boiler, tank, & shipping container mfg	0.045	0.050	90.93%
3325	Hardware mfg	0.044	0.044	77.42%
3326	Spring & wire product mfg	0.033	0.033	88.61%
3327	Machine shops, turned product, & screw, nut, & bolt mfg	0.068	0.068	96.16%
3329	Coating, engraving, heat treating, & allied activities	0.044	0.054	74.79%
3331	Other fabricated metal product mfg	0.057	0.060	53.16%
3332	Agriculture, construction, & mining machinery mfg	0.113	0.144	65.18%
3333	Commercial & service industry machinery mfg	0.069	0.074	65.91%
3334	Ventilation, heating, AC, & commercial refrigeration equip mfg	0.039	0.041	82.78%
3335	Metalworking machinery mfg	0.087	0.088	77.84%
3336	Engine, turbine, & power transmission equipment mfg	0.105	0.040	50.32%
3339	Other general purpose machinery mfg	0.071	0.075	62.42%
3341	Computer & peripheral equipment mfg	0.183	0.177	59.59%
3342	Communications equipment mfg	0.411	0.325	66.78%
3343	Audio & video equipment mfg	0.918	0.224	28.69%
3344	Semiconductor & other electronic component mfg	0.166	0.149	59.02%
3345	Navigational, measuring, medical, & control instruments mfg	0.050	0.050	72.44%
3346	Mfg & reproducing magnetic & optical media	0.053	0.064	92.99%
3351	Electric lighting equipment mfg	0.041	0.043	86.82%
3352	Household appliance mfg	0.060	0.043	79.00%
3353	Electrical equipment mfg	0.093	0.087	69.42%
3359	Other electrical equipment & component mfg	0.149	0.138	71.61%
3361	Motor vehicle mfg	0.160	0.107	72.90%
3362	Motor vehicle body & trailer mfg	0.118	0.111	87.28%

(Continued)

NAICS	Industry	U.S. Shipments Only	Total Shipments	U.S. Share
3363	Motor vehicle parts mfg	0.082	0.063	76.40%
3364	Aerospace product & parts mfg	0.096	0.079	49.06%
3365	Railroad rolling stock mfg	0.123	0.114	83.14%
3366	Ship & boat building	0.031	0.020	90.27%
3369	Other transportation equipment mfg	0.082	0.078	88.10%
3371	Household & institutional furniture & kitchen cabinet mfg	0.045	0.040	94.78%
3372	Office furniture (including fixtures) mfg	0.034	0.035	94.43%
3379	Other furniture related product mfg	0.069	0.068	98.14%
3391	Medical equipment & supplies mfg	0.029	0.021	76.21%
3399	Other miscellaneous mfg	0.069	0.031	69.88%

Source: See source note for Table 3.

End Notes

[1] Dixit and Pindyck (1994) provides an excellent introduction to the relationship between investment and uncertainty.

[2] We analyzed these specific companies because they publicly report their revenues by geographic area, separating U.S. revenues from non-U.S. revenues. This report is not intended to praise or criticize these two companies. It is simply restating information that is published in the companies' public annual reports in terms of our analytical framework.

[3] The following company profile is based on the 2009 Annual Report of the Xerox Corporation.

[4] Likewise, the reported U.S. sales are not necessarily limited to the company's U.S. production. They may include U.S. imports from the company's foreign affiliates.

[5] Throughout this paper, we report all dollar values in 2008 constant dollars.

[6] To measure volatility, we removed the industry-specific mean and linear trend from the shipments data. We discuss this adjustment to the data in the Technical Appendix. The standard deviation is the square root of the average of the squared differences between the variable and its predicted value for each industry and each year. Since there are trends in the data, the predicted values change over time.

[7] This company profile is based on the 2009 Annual Report of the Dow Chemical Company.

[8] U.S. domestic exports are produced in the United States. By definition, they include goods from U.S. Foreign Trade Zones that have been enhanced in value. They do not include re-exports. The U.S. International Trade Commission defines the free alongside ship value as "the value of exports at the U.S. port, based on the transaction price, including inland freight, insurance, and other charges. The value excludes the cost of loading the merchandise aboard the carrier and also excludes any further costs."

[9] In the Technical Appendix, we identify the technical assumptions that underlie this economic interpretation.

[10] Specifically, they imply that the covariances between the industries' U.S. shipments and their exports were negative.

[11] The NAICS codes for these five industries are 3161, 3343, 3159, 3342 and 3336 respectively.

In: United States Manufacturing
Editors: V. Murphy and J. E. Walker
ISBN: 978-1-62081-522-9

Chapter 4

REPORT TO THE PRESIDENT ON ENSURING AMERICAN LEADERSHIP IN ADVANCED MANUFACTURING*

Executive Office of the President
President's Council of Advisors on Science and Technology

ABOUT THE PRESIDENT'S COUNCIL OF ADVISORS ON SCIENCE AND TECHNOLOGY

The President's Council of Advisors on Science and Technology (PCAST) is an advisory group of the nation's leading scientists and engineers, appointed by the President to augment the science and technology advice available to him from inside the White House and from cabinet departments and other Federal agencies. PCAST is consulted about and often makes policy recommendations concerning the full range of issues where understandings from the domains of science, technology, and innovation may bear on the policy choices before the President. PCAST is administered by the White House Office of Science and Technology Policy (OSTP).

For more information about PCAST, see www.whitehouse.gov/ostp/pcast.

---

* This is an edited, reformatted and augmented version of Executive Office of the President, President's Council of Advisors on Science and Technology, dated JUNE 2 011.

The President's Council of Advisors on Science and Technology

Co-Chairs
John P. Holdren
Assistant to the President for Science and Technology
Director, Office of Science and Technology Policy

Vice-Chairs
William Press
Raymer Professor in Computer Science and Integrative Biology
University of Texas at Austin

Members
Rosina Bierbaum
Dean, School of Natural Resources and Environment
University of Michigan

Christine Cassel
President and CEO
American Board of Internal Medicine

Christopher Chyba
Professor, Astrophysical Sciences and
International Affairs Director,
Program on Science and Global Security
Princeton University

S. James Gates, Jr.
John S. Toll Professor of Physics Director,
Center for String and Particle Theory
University of Maryland, College Park

Eric Lander
President
Broad Institute of Harvard and MIT

Maxine Savitz
Vice President
National Academy of Engineering

Shirley Ann Jackson
President
Rensselaer Polytechnic Institute

Richard C. Levin
President
Yale University

Chad Mirkin
Rathmann Professor, Chemistry,
Materials Science and Engineering,
Chemical and Biological Engineering and Medicine
Director, International Institute for Nanotechnology
Northwestern University

Mario Molina
Professor, Chemistry and Biochemistry
University of California, San Diego
Professor, Center for Atmospheric Sciences Scripps
Institution of Oceanography Director,
Mario Molina Center for Energy and Environment,
Mexico City

Ernest J. Moniz
Cecil and Ida Green Professor of Physics and Engineering Systems
Director, MIT's Energy Initiative
Massachusetts Institute of Technology

Craig Mundie
Chief Research and Strategy Officer Microsoft Corporation

Ed Penhoet
Director, Alta Partners
Professor Emeritus of Biochemistry and Public Health
University of California, Berkeley

Barbara Schaal
Mary-Dell Chilton Distinguished Professor of Biology,
Washington University, St. Louis
Vice President, National Academy of Sciences

Eric Schmidt
Executive Chairman Google, Inc.

Daniel Schrag
Sturgis Hooper Professor of Geology Professor,
Environmental Science and Engineering
Director, Harvard University Center for the Environment
Harvard University

David E. Shaw
Chief Scientist, D.E. Shaw Research Senior Research Fellow,
Center for Computational Biology and
Bioinformatics Columbia University

Ahmed Zewail
Linus Pauling Professor of Chemistry and Physics
Director, Physical Biology Center
California Institute of Technology

Staff

Deborah D. Stine
Executive Director

Sridhar Kota
Assistant Director, Advanced Manufacturing and
American Society of Mechanical Engineers Fellow, OSTP

Arun Seraphin
Assistant Director, Defense Programs, OSTP

T.J. Augustine
Student Volunteer, OSTP

Writer

Bina Venkataraman
Senior Science Policy Adviser Broad
Institute of MIT & Harvard

Executive Office of the President President's Council of Advisors on Science and Technology, Washington, D.C. 20502

President Barack Obama
The White House
Washington, D.C.
20502

Dear Mr. President,

It is our pleasure to present to you this report, *Ensuring American Leadership in Advanced Manufacturing*, prepared for you by the President's Council of Advisors on Science and Technology (PCAST) and the President's Innovation and Technology Advisory Committee (PITAC).

This report provides a strategy and specific recommendations for revitalizing the Nation's leadership in advanced manufacturing.

In preparing this report and its recommendations, PCAST/PITAC held a workshop with leading manufacturing executives and innovation experts. We also consulted experts from your Administration, industry groups, and academia during the course of our study.

Although the U.S. has been the leading producer of manufactured goods for more than 100 years, manufacturing has for decades been declining as a share of GDP and employment.

Over the past decade, it has become clear that this decline is not limited to low-technology products, but extends to advanced technologies invented in the U.S., and is not solely due to low-wage competition. Moreover, it is increasingly apparent that technology innovation is closely tied to manufacturing knowledge. We cannot remain the world's engine of innovation without manufacturing activity.

We do not believe that the solution is **industrial policy**, in which government invests in particular companies or sectors. However, we strongly believe that the Nation requires a coherent **innovation policy** to ensure U.S. leadership support new technologies and approaches, and provide the basis for high-quality jobs for Americans in the manufacturing sector.

To ensure that the U.S. attracts manufacturing activity and remains a leader in knowledge production, we recommend the following two strategies: (1) Create a fertile environment for innovation so that the United States provides the overall best environment for business.

We believe this can be accomplished through tax and business policy, robust support for basic research, and training and education of a high-skilled workforce; and (2) Invest to overcome market failures, to ensure that new technologies and design methodologies are developed here, and that technology-based enterprises have the infrastructure to flourish here.

We recommend this be accomplished by launching an Advanced Manufacturing Initiative.

This initiative would support innovation in advanced manufacturing through applied research programs for promising new technologies, public-private partnerships around broadly-applicable and precompetitive technologies, the creation and dissemination of design methodologies for manufacturing, and shared technology infrastructure to support advances in existing manufacturing industries.

Our report and its recommendations serve the aims outlined in your Strategy for American Innovation, and build upon the initiatives in your 2012 budget proposal. It is an honor to provide our perspective on an issue of such vital importance to the U.S. economy and national security.

John P. Holdren
PCAST Co-Chair

Eric Lander
PCAST Co-Chair

Shirley Ann Jackson
PITAC Co-Chair

Eric Schmidt
PITAC Co-Chair

EXECUTIVE SUMMARY

Ensuring American Leadership in Advanced Manufacturing

The United States has long thrived as a result of its ability to manufacture goods and sell them to global markets. Manufacturing activity has supported our economic growth, leading the Nation's exports and employing millions of Americans. The manufacturing sector has also driven knowledge production and innovation in the United States, by supporting two-thirds of private sector research and development and by employing scientists, engineers, and technicians to invent new products and introduce innovations in existing industries.

The Nation's historic leadership in manufacturing, however, is at risk. Manufacturing as a share of national income has declined, as has manufacturing employment, and our leadership in producing and exporting manufactured goods is in question. The loss of U.S. leadership in manufacturing, moreover, is not limited to low-wage jobs in low-tech industries, nor is it limited to our status relative to low-wage nations. The United States is lagging behind in innovation in its manufacturing sector relative to high-wage nations such as Germany and Japan, and has relinquished leadership in high-tech industries that employ highly-skilled workers. Our trade balance in advanced technology manufactured products—long a relative strength of the United States—shifted from surplus to deficit starting in 2001,[1] and a trade deficit of $17 billion in 2003 further widened to $81 billion by 2010.[2] In addition, the United States has been steadily losing the research and development activity linked to manufacturing—and associated high-skilled jobs—to other nations, as well as our ability to compete in the manufacturing of products that were invented and innovated here—from laptop computers to flat panel displays and lithium ion batteries.

As U.S. manufacturing leadership is waning, other nations are investing heavily in growing and revitalizing their manufacturing sectors and are crafting policies to attract and retain production facilities and multinational companies within their borders. Such policies include partnerships, physical structures such as science parks or technology clusters, tax and regulatory incentives, and concentrated investment in commercialization of promising technologies. Some of these policies amount to industrial policy— making clear bets on specific firms and industries—but others support pre-competitive activities that would be regarded as within the scope of appropriate government action in the U.S.

Implications of Declining U.S. Manufacturing Leadership

The future ability of the United States to innovate and invent new products and industries, provide high quality jobs to its citizens, and ensure national security depends upon how well we support innovation and the development and use of advanced technologies for our manufacturing sector. While the United States may not be able to compete in the long run to make goods for which low-wage unskilled labor is the key input, this need not be true for sophisticated manufacturing linked to products and processes derived from scientific discovery and technological innovation. There are three compelling reasons why we should strive to revitalize our leadership in manufacturing:

1) Manufacturing, based on new technologies including high-precision tools and advanced materials, provides the opportunity for high-quality, good-paying jobs for American workers;
2) A strong manufacturing sector that adapts to and develops new technologies is vital to ensure ongoing U.S. leadership in innovation, because of the synergies created by locating production processes and design processes near to each other; and
3) Domestic manufacturing capabilities using advanced technologies and techniques are vital to national security.

PCAST focuses in this report on **advanced manufacturing**, a family of activities that (a) depend on the use and coordination of information, automation, computation, software, sensing, and networking, and/or (b) make use of cutting edge materials and emerging capabilities enabled by the physical and biological sciences, for example nanotechnology, chemistry, and biology. This involves both new ways to manufacture existing products, and especially the manufacture of new products emerging from new advanced technologies. We believe that advanced manufacturing provides the path forward to revitalizing U.S. leadership in manufacturing, and will best support economic productivity and ongoing knowledge production and innovation in the Nation.

The Need for an Innovation Policy

While the United States should avoid industrial policy—making bets on particular companies and industries—we should be unabashed in pursuing an innovation policy. Specifically, the Nation requires a strategy for supporting innovation in advanced manufacturing.

Summary of Key Conclusion

- The United States is losing leadership in manufacturing—not just in low-tech industries and products and not just due to low-wages abroad. We are losing ground in the production of high-tech products, including those resulting from U.S. innovation and inventions, and in manufacturing-associated research and development (R&D).
- As U.S. leadership in manufacturing declines, other nations are investing heavily in advancing their manufacturing leadership, innovation systems, and R&D.
- Advanced manufacturing has the potential to create and retain high-quality jobs in the United States.
- The Nation's long-term ability to innovate and compete in the global economy greatly benefits from co-location of manufacturing and manufacturing-related R&D activities in the United States. The loss of these activities will undermine our capacity to invent, innovate, and compete in global markets.
- A strong advanced manufacturing sector is essential to national security.
- The United States lags behind competitor nations in providing the business environment and skilled workforce needed for advanced manufacturing.
- Federal investments in new technologies, shared infrastructure, and design tools have been crucial to the birth and growth of major new industries.
- Individual companies cannot justify the investment required to fully develop many important new technologies or to create the full infrastructure to support advanced manufacturing. Private investment must be complemented by public investment. Key opportunities to overcome market failures include investing in the advancement of new technologies with transformative potential, supporting shared infrastructure, and accelerating the manufacturing process through targeted support for new methods and approaches.

The objectives of an innovation policy should be to ensure (i) that the U.S. provides the best overall environment in which to do business, (ii) that powerful new technologies are developed here and (iii) that technology-based enterprises have the infrastructure required to flourish here.

A U.S. innovation policy should include creating a business and tax environment that attracts and retains firms that invest in knowledge production and manufacture innovative products here. This can be done via the use of tailored incentives and through improved education and training of our workforce to use and develop advanced technologies.

A U.S. innovation policy should also involve building on our Nation's tradition of making strategic co-investments in precompetitive technologies that face market failure but that are critical to innovation in manufacturing. These investments should include support for new technologies that would form the basis of new industries, as well as shared infrastructure facilities that could be accessed by small and medium-sized firms for widespread benefit across industries.

Strategy & Recommendations to Ensure U.S. Leadership in Advanced Manufacturing

An overarching strategy to revitalize U.S leadership in advanced manufacturing should involve the following two components:

1. Invest to overcome market failures, to ensure new technologies are developed here and technology-based enterprises have the infrastructure to flourish here. Specifically, the Federal Government should do this by:

- Supporting applied research programs in new technologies with the potential for transforming impact
- Co-investing in public-private partnerships (PPPs) to facilitate development of broadly-applicable technologies with transformative potential
- Supporting the creation and dissemination of powerful design methodologies that dramatically expand the ability of entrepreneurs to design products and processes, and
- Investing in shared technology infrastructure that would help U.S. companies improve their manufacturing.

When co-investing to overcome market failures for key technologies, the following criteria should be applied: – The technology area has a high potential payoff in employment and output.

- There is a prospect of sustainable competitive advantage for the U.S., including through first-mover advantage.
- Identifiable market failures impede adequate private investment.
- PPPs include industrial partners are willing to co-invest with the government.
- PPPs include some industrial partners with sufficient size to invest at scale in the fruits of the pre-commercial research, as well small and start-up enterprises.
- Investments will help anchor subsequent manufacturing in the United States—for example, through shared labs, pilot plants, technology infrastructure and creation of clusters.
- Shared infrastructure will help existing firms and industries compete globally by increasing the quality and performance of their products

2. Create a fertile environment for innovation here, by:

- Encouraging firms to locate R&D and manufacturing activities in the U.S. through tax and business policies
- Supporting a robust basic research enterprise
- Ensuring a supply of skilled workers, through policies that cultivate and attract high-skilled talent

The key recommendation in this report is that the Federal Government launch an Advanced Manufacturing Initiative (AMI). We recommend that AMI be a concerted, whole-of-government effort, spearheaded by the Department of Commerce, Department of Defense, and Department of Energy and coordinated by the Executive Office of the President (EOP), either through the Office of Science and Technology Policy, National Economic Council, or the office of the Assistant to the President for Manufacturing Policy. The Secretaries of Commerce, Defense, and Energy should assign lead responsibility to an appropriate agency or agencies within the Department—such as National Institute of Standards and Technology (NIST) at Commerce, DARPA at Defense, and ARPA-E or EERE at Energy. It is crucial that this whole-of-government effort be complemented by parallel initiatives in the industry and academia. AMI should develop mechanisms to involve these sectors and to draw on their expertise in identifying technological opportunities. An external advisory board that has access to advanced manufacturing expertise should help guide this work.

The coordinating body of AMI should prepare a biennial report to the President on the most important needs for Federal investments to propel advanced manufacturing in the U.S., including (i) Coordinated Federal support to academia and industry for applied research on new technologies and design methodologies, (ii) Public-private partnerships to advance such technologies through pre-competitive consortia that tackle major-cross-cutting challenges, (iii) Development and dissemination of design methodologies that dramatically decrease the time and lower the barrier for entrepreneurs to make products, (iv) Shared facilities and infrastructure to help small and medium-sized firms improve their products to compete globally. The report should also identify the most pressing technological challenges that merit focused attention for these activities.

AMI should also report on the availability of financing for pilot plants and early-stage activities within these technology areas, and should include an analysis of comparable financing opportunities in other countries and options for providing revenue-neutral financing.

Funds to implement the programs recommended by AMI should be appropriated to the Departments of Commerce, Defense, and Energy to support the most promising opportunities. The funding level should initially be $500 million per year to be allocated across the three agencies as appropriate, rising to $1 billion over four years. Some of these funds may be drawn from existing programs as appropriate.

Below we summarize all of our specific recommendations. Further details can be found in the report.

Summary of Recommendations

Recommendation 1: Launch the Advanced Manufacturing Initiative

The Federal Government should launch an Advanced Manufacturing Initiative for America's Future (AMI). AMI should be a concerted, whole-of-government effort, spearheaded by the Department of Commerce, Department of Defense, and Department of Energy and coordinated by the Executive Office of the President (EOP).

The coordinating body of AMI should prepare a biennial report to the President on the most important needs for Federal investments, including:

- Coordinated Federal support to academia and industry for applied research on new technologies and design methodologies
- Public-private partnerships (PPPs) to advance such technologies through pre-competitive consortia that tackle major-cross-cutting challenges
- Development and dissemination of design methodologies that dramatically decrease the time and lower the barrier for entrepreneurs to make products
- Shared facilities and infrastructure to help small and medium-sized firms improve their products to compete globally.

The report should also identify the most pressing technological challenges that merit focused attention for these activities.

AMI should also report on the availability of financing for pilot plants and early-stage activities within these technology areas. It is crucial that this whole-of-government effort be complemented by parallel initiatives in the industry and academia.

AMI should develop mechanisms to involve these sectors and to draw on their expertise in identifying technological opportunities. An external advisory board that has access to advanced manufacturing expertise should help guide this work.

Funds to implement the programs recommended by AMI should be appropriated to the Departments of Commerce, Defense, and Energy to support the most promising opportunities, at the level of $500 million rising to $1 billion over four years. Some of these funds may be drawn from existing programs as appropriate. AMI should work closely with industry and academia in identifying opportunities through an appropriate advisory board.

Recommendation 2: Improve Tax Policy

The Federal Government should:

- Reform corporate income taxes, to bring the marginal tax rate in line with other OECD countries, as advocated by President Obama in his 2011 State of the Union address

- Extend the R&D tax credit permanently and increase the rate to 17%, as advocated in the Presidents' Strategy for American Innovation and FY2012 budget request.

Recommendation 3: Support Research, Education, and Workforce Training:

To ensure the health of the research enterprise that underpins innovation and national, and to ensure that the Nation has the highly skilled workforce needed to attract and maintain advanced manufacturing in the United States, the Federal Government should:

- Fulfill the President's plan to double the research budgets of three key science agencies over the next ten years: the National Science Foundation, the Department of Energy's Office of Science, and the National Institutes of Standards and Technology. Ensure appropriate research budget levels for other research agencies.
- Help fulfill the President's goal that public and private investment R&D reach 3% of GDP.
- Strengthen science, technology, engineering and mathematics (STEM) education.
- Expand the number of high-skilled foreign workers that may be employed by U.S. companies.

I. CURRENT STATE OF U.S. MANUFACTURING

The United States has long thrived because of its ability to make things and sell them in global markets. Our Nation's leadership in manufacturing has provided a foundation for economic growth, and has employed a large, though diminishing, portion of the work force. The United States was[3] the world's leading producer of manufactured goods from 1895 through 2009; some experts estimate that China surpassed the United States as the leading manufacturing country last year. The manufacturing sector continues to be a mainstay of our economic productivity, generating nearly $1.6 trillion in GDP in 2009 (11.2% of total U.S. GDP).[4] U.S. manufacturing firms lead the Nation in exports: The $1.1 trillion of manufactured goods shipped abroad constituted 86% of all U.S. goods exported in 2010.[5] The manufacturing sector employed 11.5 million workers in 2010, or 9% of total employment,[6] and supported additional non-manufacturing jobs up and down the supply chain as well as in financial services.[7] Manufacturing has also served as an engine for innovation and knowledge production. Historically, the manufacturing sector has been tightly linked with the nation's R&D activities. Manufacturing firms perform almost two-thirds of all private-sector R&D.[8] Manufacturing companies located in the United States that performed or funded R&D domestically or overseas employed an estimated 16.3 million workers in 2008, and 1.1 million of these were R&D workers (engineers, scientists, technicians and support staff).[9] Manufacturing is also important in the support of our national and homeland security. The 2010 Quadrennial Defense Review highlighted this, stating "In the mid to long term, it is imperative that we have a robust industrial base with sufficient manufacturing capability and capacity to preserve our technological edge and provide for the reset and recapitalization of our force."[10] Despite this historic strength, the U.S. manufacturing sector faces enormous challenges, and American leadership and competitiveness in manufacturing is at risk. As a fraction of U.S. GDP, manufacturing declined from 27% in 1957 to about 11% by 2009.[11] Manufacturing employment declined from 17.6 million jobs in 1998 to just 11.6 million jobs at the end of 2010.[12] For decades, we have seen the movement offshore of production facilities, a trend that began with furniture, clothing, and textiles, and has expanded to many other commodities.[13] A parallel trend is that average weekly wages in the United States have more or less remained unchanged since 1980, signaling that standards of living have not risen for most Americans over the past 30 years.[14] The loss of U.S. manufacturing leadership and jobs is not solely an issue of lower relative labor costs abroad. Strong evidence that labor costs are not the key factor is that the United States is lagging behind in driving innovation in manufacturing not just relative to low-wage nations, but also relative to nations such as Germany and Japan. Both countries make high-quality goods that command premiums in the marketplace. In 2003, Germany surpassed the United States in total export value, with key exports in machinery, vehicles, chemicals, and metals and manufactures. (Although in 2009, China surpassed Germany to become the world's leading exporter, Germany still leads the United States.)[15] Moreover, the skills and talents of workers—rather than the cost of labor—appears to matter most to companies deciding where to locate their manufacturing operations: The 2010 Global Manufacturing Competitiveness Index, a study based on the input of 400 CEOs and senior manufacturing executives worldwide, showed access to talented workers was the major driver of a country's competitiveness in attracting manufacturing, above the cost of labor and materials. The report also predicted the U.S. would slip in this competitiveness ranking by 2015.[16] This is despite the fact that BLS data showed in 2007 that U.S. compensation for manufacturing workers was lower than the average for Europe and the rate for 16 other countries.[17] The loss of U.S. manufacturing jobs is not just limited to commodities, or "low-tech" products. The trend of production migrating abroad has expanded to high-tech manufacturing: The Nation's share of the global market of exports from high-technology industries declined from around 20% in the late 1990s to about 11% in 2008.[18] The trade balance in advanced technology manufactured products—long a relative strength of the United States—shifted from surplus to deficit starting in 2001, [19] and a trade deficit of $17 billion in 2003 further widened to $81 billion by 2010.[20] [See **Figures 1 & 2**.] At the same time, China's global trade position in high-technology products moved to surplus starting in 2001, and increased from less than $13 billion in 2003 to almost $130 billion in 2008—led by trade in information and communications goods.[21] We have not simply lost low-value jobs, such as assembly, in the high-tech sector, but sophisticated engineering and advanced manufacturing activities. In addition, we are losing the higher value jobs in software and services. The outsourcing of software development to Indian companies illustrates this progression. At first, companies outsourced basic code-writing projects to Indian firms with lower costs. Now, Indian companies and workers are writing sophisticated firmware, having developed software engineering capabilities.[22]

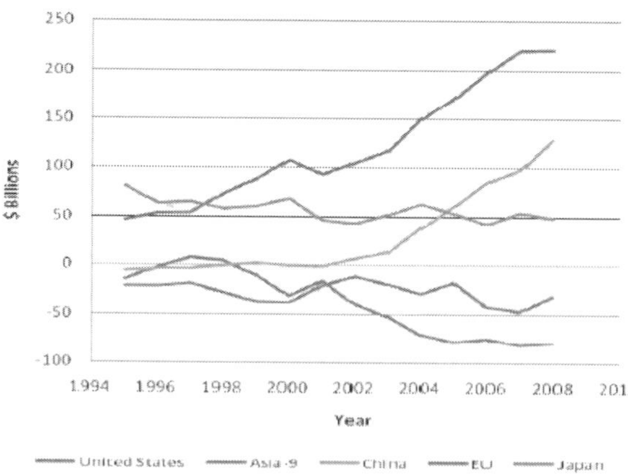

Source: National Science Board. 2010. Science and Engineering Indicators 2010. Arlington, VA: National Science Foundation (NSB 1001), Figure O34.

Trade Balance in high-technology goods for selected regions/countries: 1995-2008.

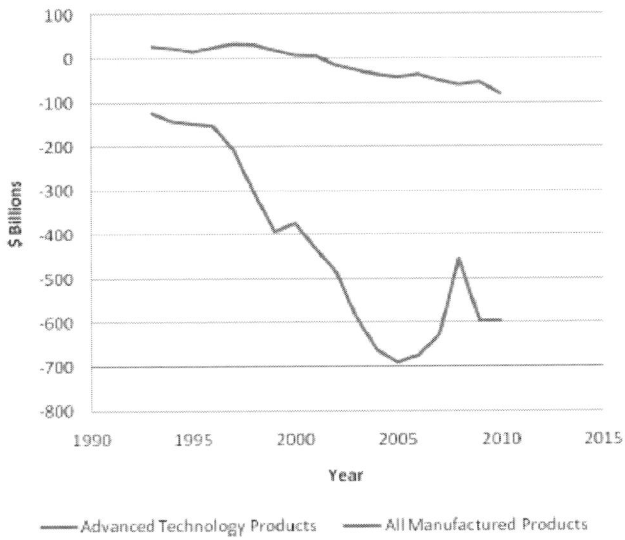

Source: Census Bureau, Foreign Trade Statistics, FT900: U.S. International Trade in Goods and Services, Exhibit 1s: Exports, Imports, and Balance of Goods by Selected NAICSbased Product Good (Manufacturing, total) and Exhibit 16a (Exhibit 15a for 2009 and earlier): Exports, Imports and Balance of Advanced Technology Products.

U.S. Trade Balances for High-Tech and All Manufactured Products.

We are also losing leadership in manufacturing industries based on inventions and knowledge that originated in the United States. Foreign firms now manufacture many products invented here. For example, the United States no longer has the knowledge, skilled people, and supplier infrastructure required to produce light-emitting diodes for energy-efficient illumination, components for consumer electronic products like the Kindle e-reader, or advanced displays for TVs, computers, and handheld devices such as mobile phones.[23] (See **Box 1**.) With respect to batteries, the United States had also lost its lead in manufacturing. (The recent Recovery Act provided $2.4 billion for advanced battery and electric drive component manufacturing, demonstrations, and infrastructure development, which should allow advanced batteries and components for plug-in and hybrid vehicles to be manufactured in the U.S. rather than be imported). Companies in Asia now design nearly every U.S. brand of cell phone and laptop computer, except for Apple.[24] This transfer of knowledge and manufacturing capacity may have national security implications as well, increasing the risk of counterfeit or malicious components in critical security systems. New U.S. companies continue to emerge in new technology sectors, but many keep costs down, access emerging markets and high-skilled workers, and satisfy their investors by locating their facilities abroad, usually in Asia—instead of creating jobs at home. [25]

Box 1: Lost Technologies

Research and innovation are essential, but alone they do not ensure a successful manufacturing sector. This is a sample[26] of technologies and products with both commercial and defense applications invented in the United States and now produced primarily abroad:

- Laptop computers
- Solar cells
- Semiconductor memory devices
- Semiconductor production equipment such as steppers
- Flat panel displays
- Robotics
- Interactive electronic games
- Lithium-ion batteries

The Nation's loss of manufacturing leadership is not limited to factory jobs; there are also concerns that we are losing leadership in R&D employment and investment related to manufacturing. R&D activity linked to manufacturing is moving offshore to access emerging global markets, and to respond to global competition for talent and the growing supply of scientists and engineers abroad. This is occurring as other countries are increasing their R&D intensities. Over the past several years, spending by U.S. firms on R&D outside the United States has grown at three times the rate of their domestic spending.[27] In the most recent employment statistics from the National Science Foundation's Division of Science Resources Statistics, three industries—all of them in the manufacturing sector—reported U.S. domestic R&D employment as a percentage of worldwide R&D employment as below 70%: communications equipment, semiconductor and other electronic components, and motor vehicles, trailers, and parts. In the last of these industries, domestic R&D employment was only 55% of the global total in 2008. At the same time, many of the nation's approximately 280,000 small and mid-size firms do not have the option to offshore R&D, and struggle to compete with foreign firms. Put together, these trends makes it increasingly possible to imagine that the United States might be shut out from competing altogether in certain industries as knowledge and inventions are increasingly produced abroad in addition to the products that result from them.

Conclusion

The United States is losing leadership in manufacturing—not just in low-tech industries and products and not just due to low-wages abroad. We are losing ground in the production of high-tech products, including those resulting from U.S. innovation and inventions, and in manufacturing-associated R&D.

Yet even as U.S. manufacturing leadership is waning, other nations are investing heavily in growing and revitalizing their manufacturing sectors and are crafting policies to attract and retain production facilities and multinational companies within their borders. Such policies include partnerships, physical structures such as science parks or technology clusters, tax and regulatory incentives, and concentrated investment in commercialization of promising technologies. Some of these policies amount to industrial policy— making clear bets on specific firms and industries, but others support pre-competitive activities that would be regarded as within the scope of appropriate government intervention in the U.S.

The following examples illustrate the international context:

- In terms of R&D investment as a fraction of GDP, the U.S. now ties for 7th in the world behind countries that include Korea, Japan, Switzerland, and Israel. Although the U.S. still accounts for 30% of global R&D, its share is shrinking.
- Unlike the United States, many advanced countries have national agencies that specifically promote technological innovation in their domestic industries, including Denmark, Finland, France, Iceland, Ireland, Japan, the Netherlands, New Zealand, Norway, South Korea, Spain, Sweden, Switzerland, and the United Kingdom. (All these nations also have agencies similar to the National Science Foundation in the U.S. that fund basic research at universities and national laboratories.) The budgets of these national innovation agencies vary widely, but Finland's Tekes invests 560 million USD annually and

Japan's New Energy and Industrial Technology Development Organization spends 2 billion USD. [28]

- Other nations are providing far more widespread and ready access to low-cost capital to innovative startup companies, through development banks that finance emerging manufacturing technologies in the clean energy sector and beyond. The China Development Bank, for example, agreed to lend 35.4 billion USD to Chinese wind and solar companies in 2010, compared to the United States' provision of $4 billion in grants and $16 billion in loan guarantees.[29] (The Bank reaped higher profits than Morgan Stanley that year.) China's government bank investment was matched by 54.4 billion USD in state and private investment.
- Even as China becomes increasingly competitive in making products and using technologies invented in the U.S., it is also producing 300,000 invention patents and about an equal number of "utility-model patents" annually[3°] and aims to reach two million in total patent counts by 2015.[31]
- The Industrial Technology Research Institute, Taiwan's major R&D agency, supports 7 research labs and 6 research centers in the country. Through ITRI, the government covers about a fourth of the cost of private sector research in new technology areas. Unlike any U.S. agency, ITRI has the sole mission of developing technology for commercialization purposes. In 2004, it had a budget of $579 million, had spun off more than 100 companies, and had sent 17,000 of its former employees (trained in science and technology) into the Taiwanese workforce.[32]
- Via the Fraunhofer Institutes, Germany's federal and state governments are co-investing with industry in applied manufacturing research. These include private-public partnerships in advanced materials, factory operation and automation, manufacturing and engineering automation, and machine tools and forming technology. The Fraunhofer Production group, which supports adaptive, digital, and high-performance production, has an operational budget of $195 million USD per year.[33]
- Despite the success of U.S. efforts in nanotechnology, which has wide potential for strengthening manufacturing industries, our share of worldwide R&D activity in nanotechnology is declining. [34]
- The Information Technology and Innovation Foundation ranked the U.S. economy last among 40 economies in terms of trends toward

improvement in innovative capacity, using a range of criteria including the change over the past decade in corporate R&D, IT investments, foreign direct investment, business climate, and corporate taxes. [35]

Conclusion

As U.S. leadership in manufacturing declines, other nations are investing heavily in advancing their manufacturing leadership, innovation systems, and R&D.

II. IMPLICATIONS OF DECLINING U.S. LEADERSHIP IN MANUFACTURING

A longstanding debate among economists and policymakers in the United States has centered on whether it matters that we are losing manufacturing industries and jobs. The relative size of the manufacturing sector has decreased substantially in every advanced economy. Some experts observe that it has been the natural course of economic development, as incomes rise and productivity increases, for jobs to migrate from agriculture to manufacturing and from manufacturing to services. In the future, it may be inevitable that the United States will not be able to compete with low-wage nations to make goods for which unskilled labor is the key input. This need not be true for sophisticated manufacturing linked to products and processes derived from scientific discovery and technological innovation. There are three compelling reasons why we should strive to revitalize our leadership in manufacturing:

1) Manufacturing, based on new technologies including high-precision tools and advanced materials, provides the opportunity for high-quality, good-paying jobs for American workers;
2) A strong manufacturing sector that adapts to and develops new technologies is vital to ensure ongoing U.S. leadership in innovation, because of the synergies created by locating production processes and design processes near to each other; and
3) Domestic manufacturing capabilities using advanced technologies and techniques are vital to national security.

In this report, we focus in particular on advanced manufacturing, which we believe offers the path forward for revitalizing manufacturing in the United States. The term refers to a family of activities that (a) depend on the use and coordination of information, automation, computation, software, sensing, and networking, and/or (b) make use of cutting edge materials and emerging capabilities enabled by the physical and biological sciences, for example nanotechnology, chemistry, and biology. This involves both new ways to manufacture existing products, and especially the manufacture of new products emerging from new advanced technologies.

Advanced Manufacturing Can Provide High-Quality Jobs

Manufacturing already provides good jobs to many American workers, and it has the potential to provide better jobs. Total hourly compensation in the manufacturing sector averages about 22% higher than average compensation in service industries.[36] The "President's Framework for Revitalizing American Manufacturing" notes, however, that wage premiums in manufacturing have been declining for workers with less formal education. Nevertheless, high-technology workers on average earn 50 to 100% more than the average of workers in all other fields.[37] By fostering a more robust high-tech manufacturing sector and better training our workers, the United States can capture the high-value added jobs involved in the manufacturing of products using advanced technologies and processes. Companies that are embracing advanced manufacturing techniques and tools, as well as those producing products based on emerging technology, are demonstrating this potential. They are proving that U.S. companies can compete when their products are high-quality, specialized, responsive to customers' needs, and made by workers whose productivity is enhanced through technology and training. Examples include AK Steel, which has flourished in the Rust Belt despite the economic downturn by employing more than 1,300 people to make customized electrical steel for domestic and export markets,[38] and General Electric, which is creating 4,000 manufacturing jobs domestically to make products including energy-efficient washers and dryers, environmental coatings, fluorescent light bulbs, sodium batteries, and jet engines.[39] G.E. is relying on high-tech machinery, skilled workers, and composite materials to create value-added parts for fuel-efficient jet engines in the United States. It is also taking advantage of state and local tax credits, and **automation**, to make energy-efficient washers and dryers in Kentucky. Smaller companies, including

Farouk Systems, Inc., a $1 billion hand-held appliance maker, and Emerson, an electrical equipment maker based in St. Louis, have also shifted some production from Asia to the United States to improve quality control and to better access their customers, relying on automation and reduced delivery distance to improve their cost competitiveness.[40] These companies currently represent exceptions to the broader trend of off-shoring of manufacturing. Further examples suggest that an advanced manufacturing sector can provide a significant number of good jobs. A recent study showed that in New England, there is unrealized potential in advanced manufacturing (defined in the study as the capability to use advanced technologies as the basis of new manufacturing or to improve processes, including precision machining, complex electronics assembly, tooling, prototyping, and engineering-manufacturing collaboration). The report estimated that between 7,500 and 8,500 jobs with the average salary of $80,000 could be created each year in the advanced manufacturing sector in New England if several barriers could be overcome.[4'] The barriers include a need for better tax incentives; collaborations across industries, firms, and with government; and better trained workers. Even in Michigan, where overall manufacturing employment has plummeted by more than a third since 2001, the advanced manufacturing sector appears promising. A study of employment in Michigan estimated that 65% (381,000) of the state's manufacturing jobs in 2007 were in the advanced manufacturing sector.[42,43] In 2009, this percent increased to 72% even as the number of manufacturing jobs fell in the state; the state lost advanced manufacturing jobs at a lower rate than it lost manufacturing jobs overall.[44]

Conclusion

Advanced manufacturing has the potential to create and retain high-quality jobs in the United States.

Advanced Manufacturing in the U.S. Will Strengthen Innovation in the U.S.

Aside from providing jobs, a strong manufacturing sector is essential if the United States is to remain the world's leader in knowledge production and innovation. Given that manufacturing companies are responsible for about 70% of industrial R&D in the United States, and employ 63% of domestic scientists and engineers,[45] the increasing movement of manufacturing abroad has a direct effect on the innovative capacity of the nation. Furthermore, when manufacturing migrates offshore so do the knowledge and capabilities that help spur innovative new technologies and allow a country to compete in new industries. Technology and innovation often follow production as it shifts abroad; other countries then gain the knowledge to capture and integrate more of the value in the global supply chain by learning to design products, and then use that knowledge, coupled with R&D, to leapfrog to next generation technologies. When this happens, companies abroad can take advantage of the synergies between design and production to capture market share for new, emerging technologies related to the production of earlier generation technologies. This is the reason, for example, according to Gary Pisano of the Harvard Business School, why the United States lags behind other nations in advanced battery technology for fuel-efficient vehicles. The consumer electronics industry, which shifted to Asia decades ago, demanded and generated the innovations that made batteries lighter and more efficient, and in doing so allowed the nations with that industry to foster manufacturing capabilities for lightweight and rechargeable batteries, including lithium ion batteries. Companies based in countries with those capabilities were better positioned to take advantage of the new market in batteries for electric and hybrid vehicles. Similarly, the capabilities in silicon-processing and thin-film coating that countries such as China, Japan, Korea, and Taiwan developed when semiconductor foundries and flat-panel display manufacturing moved to those countries have helped them to establish leadership in solar panel innovation and manufacturing. The U.S. is not a major player in solar panel technology, in part because of our lack of domestic infrastructure for thin-film and electronics manufacturing.[46] Proximity is important in fostering innovation. When different aspects of manufacturing—from R&D to production to customer delivery—are located in the same region, they breed efficiencies in knowledge transfer that allow new technologies to develop and businesses to innovate. Historically, the co-location of manufacturing and product design has been vital. Close, rapid feedback between design and production[47] allows ideas and prototypes to be quickly tested and scaled up, and allows production needs and processes to inform design. Innovation is also aided by interaction between engineers and people working in fabrication. This is because much of the knowledge underlying emerging technologies is tacit, scholars point out, requiring person-to-person contact for efficient information transfer.[48] In modern science-based industries in particular, which are multidisciplinary, innovation requires a range of skilled people from different backgrounds who can communicate effectively. Despite the cross-border, real-time exchanges that information technology has enabled, proximity still encourages people to exchange the knowledge most critical to innovation. For example, engineers in Silicon Valley are more likely to exchange more ideas with other engineers in their region than with engineers on the East Coast.[49] Harvard professors Gary Pisano and Willy Shih summed up[50] the connection between a nation's loss of manufacturing and its loss of innovative capabilities:

> In reality, there are relatively few high-tech industries where the manufacturing process is not a factor in developing new—especially radically new—products. That's because in most of these industries product and process innovation are intertwined. So the decline of manufacturing in a region sets off a chain reaction. Once manufacturing is outsourced, process-engineering expertise can't be maintained, since it depends on daily interactions with manufacturing. Without process-engineering capabilities, companies find it increasingly difficult to conduct advanced research on next-generation process technologies. Without the ability to develop such new processes, they find they can no longer develop new products. In the long term, then, an economy that lacks an infrastructure for advanced process engineering and manufacturing will lose its ability to innovate.

Our nation's ability to the produce the next generation of inventions and innovations will be greatly enhanced if companies continue to make products here, and to perform R&D here.

Conclusion

The Nation's long-term ability to innovate and compete in the global economy greatly benefits from co-location of manufacturing and manufacturing-related R&D activities in the United States. The loss of these activities will undermine our capacity to invent, innovate, and compete in global markets.

Advanced Manufacturing is a Critical National Security Capability

While the national security of the United States benefits from access to many products that are manufactured abroad, ranging from raw materials to technologically advanced components, the irreversible trend toward globalization of supply chains brings with it risks that must be countered by preparedness and appropriate hedging actions. In this regard, a vital advanced manufacturing sector is critical for national defense and homeland security.

A useful perspective on the issue is given in a 2006 National Research Council study[51] commissioned by the Defense Intelligence Agency. That study noted that most scenarios for future conflict involve military action using forces available at the time of the decision to go to war, providing little opportunity for a foreign source to deny necessary material and thereby significantly impact operations, on a tactical timescale. However, on a strategic timescale, the situation is quite different:

> "If the United States were to become strategically dependent on a foreign industrial base for items that are critical or for which the regeneration of a U.S. industrial base would take a long time, the risk would be unacceptable.... [This] possibility should be taken into account when determining what the U.S. industrial base needs to be for defense purposes."

The 2006 committee cites information technology, nanotechnology, and biotechnology as areas worthy of continued vigilance against the possibility of strategic denial. The capabilities for advanced manufacturing, as described in the present report, are precisely the types of technologies that, if ever lost by the U.S., would take an unacceptably long time to regenerate—if they could be regenerated at all. Thus does advanced manufacturing rise to the level of a critical defense capability. Furthermore, even in cases that the supply chain risk of foreign suppliers of specific material can be managed (for example, by maintaining multiple suppliers in geopolitically diverse countries), it may be impossible for us to assess risk accurately without the hands-on understanding that can be provided by domestic suppliers of (at least closely related) products. When manufacturing is outsourced, it diminishes not just the capability of producing high-tech goods, but also drains the nation's source of knowledge and talent that would be needed for the reconstitution of domestic capacity, or for a surge capability in time of strategic need.[52] Because advanced manufacturing will often involve innovative new ideas generated by smaller companies, its advance brings the possibility of broadening the Department of Defense contractor base, yielding more competition and better acquisition outcomes. According to the Department of Defense, past manufacturing programs have been shown to reduce acquisition costs. For example, in a report to Congress, DOD states that implementations of technologies and processes developed by roughly $700 million in manufacturing R&D investment through the Manufacturing Technology Program (ManTech) is projected to result in over $6 billion in acquisition cost avoidance.[53] In 2006, the Defense Science Board noted that "the need for a cohesive manufacturing research and development investment program is fundamental to rapid acceleration of near term technology capabilities to support warfighting operations and to long-term support for transition of revolutionary technologies." [54]

Conclusion

A strong advanced manufacturing sector is essential to national security.

III. Innovation Policy: Appropriate Roles for the Federal Government

Many other nations are advancing their manufacturing enterprises through policies that include industrial policy—that is, direct investment in or subsidies to specific firms. We do not believe that the Federal Government should play the role of a venture capitalist, making large bets on particular firms and industries. Rather, the best industries, firms, and products should thrive based on their ability to compete in the marketplace.[55] While the United States should avoid industrial policy, the Nation must have a robust innovation policy—an approach not of selective cultivation of industries or firms, but of creating the conditions that support innovation in advanced manufacturing. The objectives of an innovation policy should be to ensure (i) that the U.S. provides the best overall environment in which to do business, (ii) that powerful new technologies are developed here and (iii) that technology-based enterprises have the infrastructure required to flourish here.

Creating a Fertile Environment for Innovation

In the face of abundant global opportunities for the location of both manufacturing and R&D operations, firms seek access to highly trained workers, predictable and favorable regulatory and tax environments, and resources that contribute to innovation.

As other nations invest heavily in attracting manufacturing firms, educating their workforce, strengthening their industries, and promoting innovation, some are providing better conditions—more favorable tax and regulatory environments and better access to talent—to encourage firms to produce goods as well as locate their R&D activities within their borders.

- Many nations offer more attractive tax rates. U.S. average combined marginal corporate income tax rates (Federal, state and local) are 39.21%, higher than in any other OECD nation except Japan at 39.54%. Combined taxes in Germany are 30%, France 34%, Canada 29.5%, and the United Kingdom 28%; smaller countries are much lower (e.g., Chile at 17%, Ireland at 12.5%) and the OECD mean is about 30%.[56] The repatriation of cash generated outside the United States is also subject to domestic taxation to the extent that domestic tax rates exceed foreign tax rates. Business leaders argue that this dissuades companies from putting profits to work at home.[57]
- Many nations offer a variety of tax abatements and related inducements to attract manufacturing firms and their R&D operations. The effective R&D tax credit of the United States is now ranked 17th among industrialized nations by the OECD, whereas in 1992 it was ranked first. The R&D tax credit[58] has been temporary since its introduction in 1981, requiring periodic renewal by Congress. This creates uncertainty for businesses, which depend on predictable conditions for long-range planning. President Obama has proposed expanding, simplifying, and making permanent the R&D tax credit in his budget proposal and strategy. This should be implemented as an important first step.
- Worryingly, some nations may offer a more highly-skilled workforce for manufacturing. Access to high-skilled workers, the talent driving innovation, was the single most critical factor determining a country's manufacturing competitiveness, according to a recent report[59] based on the responses of 400 CEOs and senior manufacturing executives from around the world. The report predicted that the United States was likely to decline in manufacturing competitiveness in the next 5 years due to the challenges of accessing such talent. Business leaders at PCAST's workshop reported that US companies are increasingly going abroad because countries such as India, South Korea and China

provide the high-skilled workers that they need to advance their R&D enterprise, and to run the advanced manufacturing facilities that produce high-quality, customized products. Business leaders in the United States frequently expressed the view that the nation has a shortage of workers with the skills and knowledge of science, technology, engineering and mathematics (STEM) needed for advanced manufacturing. Business leaders also are concerned that stringent quotas on skilled workers from abroad under the H1-B visa program severely constrain their access to foreign talent.

Conclusion

The United States lags behind competitor nations in providing the business environment and skilled workforce needed for advanced manufacturing.

Overcoming Market Failures: Role of U.S. Investment

A major justification for government investments in science and technology is to overcome market failures. Market failures occur when private investors invest less in technology than the socially optimal level, because they cannot reap the full benefits of their investment.

New technologies frequently face market failures. The benefits of early R&D cannot be fully captured by the firm making the investment; instead, many firms or industries benefit from knowledge spillovers. Ironically, the new technologies that have high potential for growth and job creation and that are broadly applicable across several industries, also are technologies where individual private firms often under-invest most (relative to the social optimum) because the benefits of those investments do not sufficiently accrue to them in particular, but are shared widely across many firms and industries. For this reason, public investment has been critical in the early stages of R&D.

Historically, Federal technology investment—supporting basic research, funding proof of concept and promoting early commercialization through procurement—has been crucial to the creation of many technologies that have created new industries in the United States. Such investments, commonly supported by the Department of Defense, Department of Energy, National Institutes of Health, NASA, and National Science Foundation, have helped spawn entire industries, hundreds of U.S. companies and millions of high-quality jobs for Americans for decades.

The Internet, for example, was the result of long-term funding from DARPA, and later by NSF. This research was focused not only on basic science, but also on development and demonstrations of technologies such as packet switching, communications protocols, and networking infrastructure. These investments were in areas where the return on invested capital would have been difficult for any single company to capture, and the investments required a long time horizon not easily envisioned by industry leaders.

Significant Federal funding—in some cases paired with industry funding—was also critical for the development of the transistor by Bell Labs in the 1950s, the growth of the semiconductor industry and the birth of Silicon Valley, and to the development of GPS. The Federal government funded in part the major corporate laboratories that laid the foundations for U.S. economic leadership and innovation in the 20th century, including Xerox PARC, RCA David Sarnoff Research Center, and AT&T Bell Labs.[60] According to one study, most of the top annual innovations in the U.S. over the past two decades were supported by a combination of government and industry funding.[61]

The Federal government has also used public-private partnerships as a vehicle to advance key technologies. The SEMATECH consortium of the late 1980s and early 1990s, for example, was a partnership between DARPA and 14 U.S.-based semiconductor manufacturers, in response to the fact that Japan had captured a large portion of the integrated circuit memory chip market, and was poised to capture the majority of that market. This was viewed as a risk to national competitiveness and national security. SEMATECH, which matched $500 million in Federal government spending with equivalent industry spending over 5 years, advanced precompetitive research on the technology needed for next-generation chips. It also funded a test integration facility for tool and equipment suppliers in the semiconductor industry, allowing for prototyping of innovations in chip technology through direct contact between suppliers and the major companies in the consortium. More recently, the Federal government and industry partners have launched a pilot public-private partnership to advance research on the practical, next-generation computing devices that can replace the conventional semiconductor chip by 2020. The partnership was launched by NIST and the Nanoelectronics Research Initiative (part of the Semiconductor Research Corporation). To date, the partnership has received only modest Federal funding, of about $2.7 million per year, which has leveraged about $20 million in co-investment from the states and industry partners to establish regional research centers. A group of 35 universities in 20 states is contributing to the initiative. With more significant Federal investment and industry co-investment, this initiative could help advance the critical precompetitive research needed to lead in the future of computing devices. Finally, we note another option that government can use to make cost-neutral investments that increase access to low-cost capital for innovative technology enterprises that serve national priorities and address societal challenges. Over the long run, private capital markets are efficient at identifying the best technologies; but there can be significant time-lags with early-stage technologies. The experience of some national development banks suggests that co-financing can encourage the transition of promising technologies from the stage of invention to pilot plants for production, while yielding returns for the government. This approach is being increasingly used by other nations.

Conclusion

Federal investments in new technologies have been crucial to the birth and growth of major new industries.

The next sections discuss three ways in which Federal Government might help overcome market failures that constrain U.S. leadership in manufacturing and the growth of the advanced manufacturing sector.

Opportunities: Advancing New Technologies

A first way in which the Federal Government can help overcome market failures is to provide support to advance specific early-stage technologies that have transformative potential, but for which rational private actors will under-invest because they cannot capture the full return on their investment, owing to spillover effects that would accrue to many industries and firms. It is important to distinguish support for specific technology areas versus support for specific technology companies. We do not propose that the Federal Government should play the role of venture capitalist. However, there are appropriate steps that the Government can take to advance key areas, including supporting applied research in promising technologies and precompetitive public-private partnerships and consortia. Such steps are economically justifiable when clear market failures limit progress in technologies with the potential for transformative impact on jobs and output, and sustainable competitive advantage for the U.S.. They are particularly appropriate when the technologies address societal challenges and technical and scientific problems of national importance. We elaborate on these approaches in Chapter 4. PCAST has not sought to define the specific technology areas that should be supported in this manner; this is better done through a process that we describe below. For the purposes of illustration, we briefly describe four examples of promising technologies that face potential market failure such as described above.

- **Nano-scale Carbon Materials:** Materials such as graphene, buckeyballs, and carbon nano-tubes that have nano-scale crystalline structures could serve markets for data storage, energy, optoelectronics, avionics, defense and packaging. Potential products include highly attuned chemical and biological sensors, fuel cells, touch screens, lightweight body armor, and airframes. Their properties could also allow the materials to replace silicon for next-generation integrated circuits, and to be applied to uses not yet conceivable. Because these materials could serve a diverse range of industries including energy, information storage, automotive, aerospace and microelectronics, it is hard for individual firms—or even individual industries—to reap the benefits of high-risk investments to realize the full potential of the technologies.
- **Next-generation optoelectronics:** Technology that integrates electronics and photonics to convert light into electrons, and vice versa, currently enables many products and services in telecommunications, entertainment, e-commerce and medicine. The next generation[62] of the technology requires development and production of vastly more integrated, higher-performance, and lower cost devices and systems, which could also allow for the emergence of new microelectronics, order-of-magnitude improvements in data transfer rates within and between computers without increased cost, biosensors for home healthcare and personalized medicine, imaging and night-vision systems, and faster and more capable information synthesis and processing systems. The critical innovation needed in this field is the ability to integrate leading-edge photonic functionality into the same chips as transistor-based electronics. This will leverage the existing infrastructure for electronic silicon chip fabrication to rapidly advance the manufacturing base for new photonic systems.

The benefits of developing this technology are spread across a wide range of industry sectors and products. Individual firms cannot justify the level of investment needed to fully develop the technology.

- **Flexible electronics:** The technology that allows electronic circuits to be printed on thin, flexible substrates that can be shaped without damage has the potential to improve a range of existing products including displays and photovoltaic panels and to form the basis of new products, such as bandages that sense the presence of infection and alert medical staff, foldable solar panels for earth and space applications, clothing that monitors and displays physiological signs, phased-array antennae for distributed mobile communications networks, and ultra-efficient lighting. The research and development needed to advance this technology is costly and complex, requiring facilities for prototyping and pilot-scale manufacture. Yet the benefits are broadly applicable across industries and cannot be fully captured.
- **Nanotechnology Enabled Medical Diagnostic Devices and Therapeutics:** Advances in nanomedicine have the potential to revolutionize the pharmaceutical industry, by dramatically improving prevention, diagnosis and treatment. Recent advances in nanotechnology have led to ultrasensitive assays for identifying and treating diseases at earlier stages than conventional diagnostic tools, new ways of delivering powerful therapeutics to the point of disease within the body without unintended side effects, and the development of entire new classes of pharmaceuticals based upon nanostructures for broad classes of disease such as cardiovascular disease and cancer. The ability to design products and manufacture using this technology requires new tools and procedures, scale-up plants, and a work force with a new set of skills. The technology has the potential to create new US-based industries, lower the cost of health-care, and positively impact human life. But, individual medical firms cannot rationally justify the optimal level of investment based on the return that they can achieve.

Opportunities: Supporting Shared Infrastructure

A second way that the Federal Government can promote advanced manufacturing in the U.S. is to improve access to infrastructure, especially for small and medium sized firms. There are many tools and technologies that can improve the ability of existing firms to prototype rapidly and virtually, produce small batches, customize products to individual consumers and clients, reduce inventories, and expand the range of products that they can manufacture. Many firms, however, cannot gain access to such technologies. The minimum investments required are too large to be cost-effective for an individual firm, and there is often no effective way to buy shared services. For example, powerful computational tools and resources for modeling and simulation could allow many U.S. manufacturing firms to improve their processes, design, and fabrication. They would benefit from readily accessible shared infrastructure, providing both equipment and expertise. Infrastructure currently provided at Federal laboratories, for example, for the fabrication of micro-electromechanical systems, has allowed for new products to be developed by U.S. companies. Similarly, measurement tools for nanomaterials would be enabling for many firms.

Opportunities: Rethinking the Manufacturing Process

A third way that the Federal Government can promote advanced manufacturing in the U.S. is by supporting broad efforts to dramatically rethink the manufacturing process. We believe that emerging technologies—if accelerated, integrated, and synthesized—have the potential to change how we manufacture things ranging from vehicles to personal effects, from tools to proteins. Harnessed and driven appropriately, they have the potential to create a new manufacturing base that can move quickly, cost-effectively and seamlessly from bits to atoms and from parts to systems. The resulting manufacturing tools and resources may enable:

- development cycles at reduced risk, time and cost;
- adaptive design and development for diverse market segments; and
- orders of magnitude expansion of number and diversity in product designers.

Past efforts to accelerate product development cycles and reduce costs of transition to large scale manufacturing have often focused on rapid prototyping tools and techniques. While there have been important developments in these areas, rapid prototyping has not solved the fundamental challenges of manufacturing. Rather than making prototyping techniques faster, more functional, and closer to mass production, it may make more sense to work backward from manufacturing—by developing the tools and techniques that would enable large scale manufacturing equipment and materials to efficiently produce low volume lots—down to units of one. This would enable production of prototypes in exactly the same facility that will produce units in high volume. Prototyping new products in exactly the same high-volume manufacturing facility using the same materials and processes that will be used in the final product could revolutionize production speed and efficiency in the transitions between stages. It could eliminate the time, cost, and risk that arises from transitions between stages that include prototyping, early production runs, limited, and large-scale manufacturing. These transitions currently require extensive rework, and are the source of production delays, surprises, and cost overruns. The power of this approach has been demonstrated historically in the semiconductor and information technology industries. In semiconductors, the work of Mead and Conway in the late '70s fundamentally transformed the design of very large-scale integrated (VLSI) systems. They developed an 'abstract' set of design rules that made it possible to decouple the design of systems from the specific fabrication technology used to produce them. As a result, design and simulation tools with higher level of abstractions became possible. These abstractions and tools involved a trade-off that required some loss of optimal performance in individual components. But, they allowed much greater ease and accuracy of the overall system design. These advantages led to system-level innovations that increased system performance, reduced time and cost of system design, and increased reliability and predictability of system performance. The tools and techniques led to a greater diversity and number—from hundreds to hundreds of thousands—of VLSI designers. These designers in turn contributed to a proliferation of new semiconductor products, dozens of multi-billion dollar companies and an entire new industry sector—fabless semiconductor companies. Similar to the semiconductor sector, information technology moved from machine-level, assembly language programming in the 1960s to high-level languages that decoupled the programming from the specific microprocessor or machine that was being programmed. High-level languages involved a deliberate tradeoff of some loss of performance for ease of programming and portability of programs. As in semiconductors, the move to higher-level languages, higher levels of abstraction, also led to a greater diversity and number of programmers, which in turn led to an explosion of new programs, applications and the information technology industry. Manufacturing sectors such as electromechanical systems and pharmaceuticals may involve greater levels of complexity and technical challenge than semiconductors or programming alone. Similar approaches, however, hold promise to bring comparable growth, jobs, and entrepreneurial innovation to key industries and U.S. manufacturing excellence. We see three areas for attention. The first area is in design tools that dramatically improve the existing systems engineering, integration, and testing process for complex electromechanical, cyber-physical systems that represent the bulk of manufactured products today— from toasters to automobiles. These tools could aim to develop general methods for cyber-physical products far more complex and heterogeneous than those to which such methods are applied today. These capabilities would further enable designers to use common methods throughout the design process to optimize system performance, and to verify designs in silico. This could dramatically reduce the need for expensive build, test and design cycles. The second area would develop manufacturing facilities similar to today's semiconductor foundries. The input would be verified system designs, specified and developed with the design tools above. The systems would be capable of rapid reconfiguration to accommodate a wide range of design variation. Such foundries would compress substantially the time required to go from design to product. This could advance manufacturing toward flexible, programmable, and potentially distributed production capabilities able to accommodate a diverse range of systems and system variants, rather than requiring separate facilities for single products. The third area would support generating open-source collaboration environments for the creation of large, complex, cyber-physical systems by numerous affiliated or unaffiliated designers—with the goal of democratizing the design innovation process by engaging a vastly larger pool of talent than current industry models. The development of complex software systems, for example, has benefitted significantly from the ability to leverage crowd-sourced innovation in the form of open source code development.

Conclusion

Individual companies cannot justify the investment required to fully develop many important new technologies or to create the full infrastructure to support advanced manufacturing. Private investment must be complemented by public investment. Key opportunities to overcome market failures include investing in the advancement of new technologies with trans-formative potential, supporting shared infrastructure, and accelerating the manufacturing process through targeted support for new methods and approaches.

IV. Recommendations: Toward a Renaissance in Advanced Manufacturing

In a spirit consistent with America's past successes, we recommend the following strategy for promoting advanced manufacturing in the United States.

Principles for Promoting Advanced Manufacturing in the United States

(1) Invest to overcome market failures, to ensure new technologies are developed here and technology-based enterprises have the infrastructure to flourish here.

- Support applied research programs in most promising new technologies
- Co-invest in public-private partnerships to facilitate development of broadly-applicable technologies with transformative potential
- Develop and disseminate design methodologies that dramatically decrease the time and lower the barrier for entrepreneurs to make products
- Invest in shared technology infrastructure that would help U.S. companies improve their manufacturing

(2) Create a fertile environment for innovation here.

- Encourage firms to locate R&D and manufacturing activities in the U.S. through tax and business policies
- Support a robust basic research enterprise
- Ensure a supply of skilled workers through policies that cultivate and attract high-skilled talent

The recent passage of the America COMPETES Reauthorization Act of 2010 establishes a valuable framework for activities to support advanced manufacturing (see Appendix 1). The Act authorizes the Federal government to work with industry and national laboratories to identify regulatory and tax barriers to advanced manufacturing, identify challenges appropriately tackled by PPPs, and encourage the formation of PPPs. It authorizes an interagency committee under the NSTC to plan and coordinate federal activities on advanced manufacturing. It also directs NIST to study the barriers to small and medium sized firms' use of modeling and simulation tools and techniques, and requires the Department of Commerce to recommend actions to overcome any barriers that are identified. We urge funding be appropriated for these activities. The President's FY2012 budget request and President's Strategy for American Innovation lay out specific actions to catalyze advanced manufacturing initiatives (see Appendix 2) that include modifications to the R&D tax credit, modest R&D funding for advanced manufacturing at various agencies, foundational support for public-private partnership programs at NIST, and investments in STEM education. We also urge appropriate funding for these activities. Our goal here is to build upon this foundational work, including by identifying specific actions and the funding required.

Overcoming Market Failures: Advanced Manufacturing Initiative for America's Future

There are systematic market failures that (i) block or slow the development of important new technologies and methodologies and (ii) limit access by firms to technology infrastructure. The Federal Government has historically made visionary investments that have facilitated the birth of new technology-based industries and strengthened the development of existing industries. These investments have paid enormous financial and social returns to the Nation. It is essential that we renew this wise policy. PCAST believes there are a number of Federal investments that could have large returns in propelling advanced manufacturing. These include co-investing in the advancement of new technologies that face market failure, support of shared infrastructure, and rethinking the manufacturing process through targeted support for new methods and approaches. Currently, key technological investments in this vein are being made by the Defense Advanced Research Projects Agency (DARPA), the National Institute of Standards and Technology (NIST) and the Department of Energy. Rather than specifying precise areas for investment, however, PCAST favors the creation of a disciplined and transparent process to evaluate technology opportunities, define problems or cross-cutting challenges, and make investments. Investments should be made based on the merits of proposals from industry and academia, evaluated by the criteria discussed below.[63] Toward this end, we recommend the creation of an **Advanced Manufacturing Initiative for**

America's Future (AMI). AMI should be a concerted, whole-of-government effort, spearheaded by the Department of Commerce, Department of Defense, and Department of Energy and coordinated by the Executive Office of the President (EOP), either through the Office of Science and Technology Policy, National Economic Council, or the office of the Assistant to the President for Manufacturing Policy. It is crucial that this whole-of-government effort be complemented by parallel initiatives in the industry and academia. AMI should develop mechanisms to involve these sectors and to draw on their expertise in identifying technological opportunities. An external advisory board that has access to advanced manufacturing expertise should help guide this work. AMI should focus on two components: identifying opportunities and investing in opportunities. The first component requires no new appropriations and should be launched immediately. The second component could require new appropriations or reprioritization of existing funds.

Identifying Opportunities

The Coordinator of AMI should deliver a biennial report to the President identifying and analyzing the most promising technology areas related to advanced manufacturing where Federal investment is needed and justified. The types of investments to be considered should include:

- Coordinated Federal support to academia and industry for applied research on new technologies and design methodologies with transforming potential,
- Public-private partnerships to advance such technologies through pre-competitive consortia that tackle major-cross-cutting challenges,[64]
- Development and dissemination of design methodologies that dramatically decrease the time and lower the barrier for entrepreneurs to make products,
- Shared facilities and technology infrastructure to help small and medium-sized firms improve their products to compete globally.

The report should also identify specific challenge areas for the Nation that merit focused attention. (Possible examples of challenge areas might include: highly flexible manufacturing for biopharmaceutical production; improving photovoltaic systems to become the cheapest source of power within the decade; or design, simulation, and production of electromechanical systems with the complexity of automobiles or infantry fighting vehicles.) In addition, the report should study the availability of financing for these areas of technology, including comparison of access to low-cost financing in other nations and evaluation of options for providing revenue-neutral development investment opportunities. In addition, consideration should be given to addressing the challenges that surround manufacturing readiness and the scale-up of manufacturing processes.

Investing in Opportunities

We recommend that funding be appropriated to the Department of Commerce, the Department of Defense, and the Department of Energy (DOE) to allow NIST, DARPA and DOE to fund the most promising opportunities that will respond to cross-cutting challenges and problems defined by AMI, through programs at each agency. At NIST, the AmTech program provides a model for how this can work by providing a vehicle for public-private partnerships, with mechanisms for cost-sharing by industry and states. Currently, AmTech is only a pilot program with a proposed budget of about $12 million. Germany's Fraunhofer Program offers a potential model for shared infrastructure facilities and resources. We believe that the Federal investment level should initially be $500 million per year, to be allocated across the three agencies, rising to $1 billion over four years. Some of these funds may be drawn from existing programs as appropriate. We note that the recent reauthorization of the America COMPETES Act provides for public-private partnerships in key advanced manufacturing technologies to be formed and funded by the Federal Government, but that the implementation of PPPs depends upon funds being allocated for this purpose. (We note that the Presidents FY2012 budget request proposes funds for this purpose). The Federal investment should leverage substantial investment by industry—especially in the case of PPPs, where industry might perhaps cover half the cost. Such cost-sharing is an important indicator of industrial commitment and will help ensure that these investments pay off in terms of economic growth. We recognize that budgetary constraints will make new investments difficult in the coming years. However, we believe that the proposed program has the potential to pay large economic returns to the Nation relative to the funding required.

Criteria

Investments should be made where there is a compelling case. In the case of support for applied research in specific technologies or in new design and manufacturing methodologies, these areas should be justified by the potential for transformative impact. In the case of public co-investment in PPPs, AMI should solicit specific proposals from consortia of private and public organizations. (Where necessary and appropriate, it should offer modest-sized planning grants to support the preparation of such proposals; we believe planning grants can be funded from existing funds.) Opportunities should be selected based on the merits of the proposals, and based on the below criteria:

- The technology area has a high potential payoff in employment and output.
- There is a prospect of sustainable competitive advantage for the U.S., including through first-mover advantage.
- Identifiable market failures impede adequate private investment.
- PPPs include industrial partners willing to co-invest with the government.
- PPPs include some industrial partners with sufficient size to invest at scale in the fruits of the pre-commercial research, as well small and start-up enterprises.
- Investments will help anchor subsequent manufacturing in the United States—for example, through shared labs, pilot plants, technology infrastructure and creation of clusters.
- Shared infrastructure will help existing firms and industries compete globally by increasing the quality and performance of their products

AMI should also establish metrics for determining effectiveness of these investments made by agency-specific programs, a function that could be coordinated through the NSTC subcommittee on advanced manufacturing.

Recommendation 1: Launch Advanced Manufacturing Initiative

The Federal Government should launch an Advanced Manufacturing Initiative for America's Future (AMI). AMI should be a concerted, whole-of-government effort, spearheaded by the Department of Commerce, Department of Defense, and Department of Energy and coordinated by the Executive Office of the President (EOP), either through the Office of Science and Technology Policy, National Economic Council, or the office of the Assistant to the President for Manufacturing Policy. The Secretaries of Commerce, Defense, and Energy should assign lead responsibility to an appropriate agency or agencies within the Department—such as National Institute of Standards and Technology (NIST) at Commerce, DARPA at Defense, and ARPA-E or EERE at Energy.

It is crucial that this whole-of-government effort be complemented by parallel initiatives in the industry and academia. AMI should develop mechanisms to involve these sectors and to draw on their expertise in identifying technological opportunities. An external advisory board that has access to advanced manufacturing expertise should help guide this work.

The coordinating body of AMI should prepare a biennial report to the President on the most important needs for Federal investments to propel advanced manufacturing in the U.S., including (i) Coordinated Federal support to academia and industry for applied research on new technologies and design methodologies, (ii) Public-private partnerships (PPPs) to advance such technologies through pre-competitive consortia that tackle major-cross-cutting challenges, (iii) Development and dissemination of design methodologies that dramatically decrease the time and lower the barrier for entrepreneurs to make products, (iv) Shared facilities and infrastructure to help small and medium-sized firms improve their products to compete globally. The report should also identify the most pressing technological challenges that merit focused attention for these activities.

AMI should also report on the availability of financing for pilot plants and early-stage activities within these technology areas, and should include an analysis of comparable financing opportunities in other countries and options for providing revenue-neutral financing.

Funds to implement the programs recommended by AMI should be appropriated to the Departments of Commerce, Defense, and Energy to support the most promising opportunities. The funding level should be initially be $500 million per year to be allocated across the three agencies as appropriate, rising to $1 billion over four years. Some of these funds may be drawn from existing programs as appropriate.

Because advanced manufacturing has unique national security implications, and because DARPA has, within the DoD, a unique ability to connect to small, entrepreneurial companies, DARPA has a vital role in Recommendation 1's proposed initiative. More broadly within DoD, initiatives in manufacturing going beyond Recommendation 1 may be an important part of acquisition reform and the improvement of supply chain management. Such initiatives can help transition and mature early stage DoD research investments into technologies ready for scale up and deployment into national security systems in a more cost effective and timely fashion than available at present. AMI should engage experts from industry and academia in identifying technological opportunities, and work with an external advisory board that has access to advanced manufacturing expertise.

In implementing AMI, the Federal Government should also consider incentives and policy tools that will anchor advanced manufacturing activities to the United States.

Box 2: Examples of Important Technology Areas for Advanced Manufacturing

The following technologies represent potential areas that would align with the criteria PCAST has outlined for public-private investment to support advances in manufacturing:

Advanced robotics. Next-generation robots could be mobile and autonomous in their environment, with the ability to interact with their environment and achieve outcomes without programming of all procedures. They also have the potential to be safely operable around humans or in dangerous environments. Intelligent automation could build robots'capabilities to increase autonomy and flexibility to enable manufacturers to respond to needs and desires of customers very efficiently. Major advances could provide broad-based innovation benefiting multiple industries, such as that provided by computer-aided design and computer-aided manufacturing, total quality management, and just-in-time manufacturing. The private sector currently under-invests in robotics technology as the full benefits are dispersed across many firms and industries. Public-private investments to advance robotics technology could help US manufacturers compete by allowing quick, nimble improvements in product quality, productivity, time to market, and cost.

Nanoelectronics. Semiconductors have been one of the key enabling technologies of the information technology revolution. The technology has driven ongoing improvements in price and performance of computers, phones and other communications equipment. Experts believe that today's microprocessor technology (the silicon CMOS field effect transistor) will reach its limits in terms of performance around 2020. The most promising candidates for the next "logic device" are being identified collaboratively by experts in industry, academia, and government, through joint research initiative known as the Nanoelectronic Research Initiative (NRI). New research on fabrication techniques, materials, heat flow and power dissipation are necessary to achieve the next generation of technologies that will support a broad range of industries and applications.

The benefits of investing substantially in such research are widespread but difficult for individual industries and firms to capture alone. Federal co-investment is important to ensuring the U.S. industries leads the next microprocessor technology, as it led the last.

Materials by Design. Materials that endure in extreme temperatures, lightweight composites and new electronic and functional materials have in the past enabled advances in transportation, electronics, and aerospace and have given U.S. companies a competitive edge.

A new generation of materials is needed for a broad range of applications including energy storage, flexible electronics, and stronger, more versatile defense products.

But new materials often take 10-15 years from the initial research stage to be integrated into applications, and there is a trend toward increasing speed of product design. U.S. companies require tools for rapid discovery, development, and use of advanced materials.

Investments need to be made in developing broadly available tools. These include modeling and software tools, databases of design material properties, and tools for testing materials more rapidly.

Biomanufacturing. Researchers are developing new tools that could allow us to readily engineer biological systems, with widespread applications for energy, medicine, and electronics.

Further development of these tools could dramatically reduce the time and cost of realizing biotechnology applications such as biosynthesis of pharmaceuticals, efficient conversion of sunlight and carbon dioxide into fuel, and assembly of inorganic materials into electronic devices or batteries.

The development of enabling science and technologies is an important potential area for Federal co-investment.

Creating a Fertile Environment: Tax Policy, Research Enterprise and Skilled Workers

In addition to this strategic initiative to promote advanced manufacturing, the Nation must cultivate the conditions to attract and retain advanced manufacturing activities in the United States. This involves several facets of innovation policy:

Corporate Tax Policy

The Federal Government has an essential role in shaping the environment in which U.S. manufacturers compete, through its tax policies. We urge the Federal Government adopt two policies outlined in the President's plan, which are widely supported by the scientific and business communities.

Recommendation 2: Improve Tax Policy

The Federal Government should:

- Reform corporate income taxes, to bring the marginal tax rate in line with other OECD countries, as advocated by President Obama in his 2011 State of the Union address. This will create stronger incentives to locate manufacturing plants in the U.S. and help eliminate the disincentive to repatriate profits, and instead, draw companies to use them to employ U.S. workers.
- Extend the R&D tax credit permanently and increase the rate to 17%, as advocated in the Presidents' Strategy for American Innovation and FY2012 budget request. Knowing that the credit will persist will encourage firms to lengthen their time horizon for R&D investments. The rules governing the tax credit should also be examined to make clear that R&D on manufacturing processes qualifies for the credit.

Robust Research Enterprise

Basic and applied research is the critical foundation for innovation. As stated in the first *Gathering Storm* report from the U.S. National Academy of Sciences, "A balanced research portfolio in all fields of science and engineering research is critical to U.S. prosperity."[65] President Obama has recognized this need for a strong research base to support advanced manufacturing in the administration's framework to revitalize manufacturing. The Federal Government has traditionally supported both basic and applied scientific and engineering research, including much of the research performed by universities. Yet, the Federal Government's funding of R&D as a fraction of GDP has declined by 60% over the past 40 years.[66] Corporate R&D spending has risen rapidly, but it is directed largely toward development. The Federal Government largely retains responsibility for the funding of basic research, performed largely at the Nation's research universities and institutes.

Supply of Skilled Workers

In the long term, the strength of our innovation system depends on the skill of our workforce. Manufacturers increasingly need employees who bring substantial technological abilities. As the original *Gathering Storm* report stated, "The competitiveness of U.S. knowledge industries will be purchased largely in the K-12 classroom." Importantly, we need a workforce that includes not only scientists and engineers with advanced degrees, but also factory-floor engineers able to oversee and improve complex manufacturing processes and workers able to use sophisticated tools and machinery in factories.

Manufacturing firms frequently cite the inability to find an adequate supply of factory engineers and workers as a barrier to locating manufacturing in the U.S. Numerous studies have described the challenges in K-12 science, technology, engineering, and mathematics (STEM) education in the U.S., which include a shortage of highly qualified and effective teachers, the lack of dynamic, adaptive instructional materials, low achievement on international standardized tests and national assessments, and low student interest in STEM-related careers.[67] PCAST has explored this topic and delivered to the President a recent report on K-12 STEM education.[68] In this report, PCAST recommends specific actions to improve the effectiveness of Federal investments in K-12 education. In 2011, PCAST plans to deliver a follow-up report on undergraduate STEM education.

In addition, the United States must continue to draw the best talent from abroad. In the near term, steps must be taken to expand the number of available H1-B visas for those with advanced education in science and engineering. Due to the stringent constraints on visas, tens of thousands of highly qualified graduates of U.S. colleges and universities return to their home countries each year. U.S. companies have tremendous need for such workers.

Moreover, foreign-born scientists and technologists greatly contribute to the U.S. economy by starting enterprise here. Roughly one-third of start-up firms in Silicon Valley are started by foreigners.

Recommendation 3: Support Research, Education, & Training

To ensure the health of the research enterprise that underpins innovation and national security and to ensure that that the Nation has the highly skilled workforce needed to attract and maintain advanced manufacturing in the U.S, the Federal Government should:

- Fulfill the President's plan to double the research budgets of three key science agencies over the next ten years: the National Science Foundation, the Department of Energy's Office of Science, and the National Institutes of Standards and Technology.[69] Ensure that other research agencies have appropriate research budget levels.
- Use Federal policy and leadership to fulfill the President's goal that public and private investment R&D reach 3% of GDP.
- Strengthen science, technology, engineering and mathematics (STEM) education, as described in the recommendations in PCAST's recent report.
- Expand the number of high-skilled foreign workers that may be employed by U.S. companies. This can be done by such policies as allowing foreign students that receive a graduate degree in STEM from a U.S. university to receive a green card, allowing each employment-based visa to automatically cover a worker and his or her spouse and children[70], and increasing the number of H1B visas.

Appendix A: Advanced Manufacturing Provisions in America Competes Act

The America COMPETES Act, passed by Congress in December 2010 and signed into law by President Obama in January 2011, contains provisions that support advanced manufacturing.

- It authorizes increased funding for basic research funding through the National Science Foundation (NSF), the Department of Energy's Office of Science, and the National Institute for Standards and Technology (NIST) in the Department of Commerce (DOC).
- It authorizes policies supporting science, technology, engineering and mathematics (STEM) education, as well as policies to strengthen the nation's innovation and competitiveness.
- It directs the Department of Commerce to analyze: the innovation capacity and economic competitiveness of the United States; the performance of the U.S. relative to other nations; the business, tax, and regulatory climate for innovation; trade and export policies; workforce issues; the effectiveness of Federal, state and regional policies; and barriers to U.S. competitiveness in emerging technology

areas. In consultation with an Innovation Advisory Board of industry and other experts, the Department will develop a 10-year innovation and competitiveness strategy for the United States to address the issues in the study.

- It directs the creation of an interagency committee on technology under the National Science and Technology Council (NSTC) with responsibility to plan and coordinate Federal programs and activities on advanced manufacturing, including R&D programs of federal agencies, translation and commercialization of federally-funded technology into manufacturing, and the formation of private-public partnerships.
  It requires that the committee set and make strategic plans to achieve short and long-term objectives for R&D, worker training, and assistance to small and medium-sized manufacturers.
- It directs the Manufacturing Extension Partnership, a program of NIST, to improve training at community colleges of workers to serve the needs of small and medium-sized manufacturing businesses, and to evaluate the barriers faced by small-sized manufacturers and assist in responding to these challenges via the regional manufacturing extension centers.
  The MEP would also be expanded to support construction and green energy industries.
- It establishes a dedicated program in NSF to fund fundamental research at higher-education institutions that will lead to "transformative advances" in manufacturing technologies, processes, and enterprises. Proposed research areas include: nanomanufacturing; machines and equipment for manufacturing and construction including robotics and automation; manufacturing enterprise systems; advanced sensing and control techniques; materials processing; and information technology for manufacturing including modeling and simulation capabilities.
- It directs NSF to support higher education institutions for STEM internship programs for undergraduate students to work in industry. The NSF would also fund programs to strengthen and expand technical and scientific education and training in advanced manufacturing, including through NSF's Advanced Technological Education program.

- It directs DOC to create a program of Federal loan guarantees for small and medium—sized manufacturers pursuing the use or production of innovative technologies for manufacturing.
- It directs DOC to study barriers to use of high-end computing simulation and modeling by small-and medium-sized U.S. manufacturers, including access to facilities and resources, availability of software and technologies, and access to expertise, and tools to manage costs.

  The study would yield recommendations for responding to these challenges to increase use of high-end computing modeling and simulation. In addition, Federal agencies could carry out demonstration and pilot programs as part of the study.

APPENDIX B: PRESIDENT OBAMA'S STRATEGY FOR AMERICAN INNOVATION AND THE FY 2012 BUDGET REQUEST: IMPLICATIONS FOR ADVANCED MANUFACTURING

The President's Strategy for American Innovation, updated in early 2011, and the President's 2012 budget propose several initiatives and programs to support advanced manufacturing.

These programs provide a strong foundation for a broader strategy on Advanced Manufacturing, as described in our recommendations. The provisions include the following:

- Increased R&D budget for advanced manufacturing technologies: Funding for the NSF, NIST, DOE and DARPA intended to support innovation to reinvigorate existing manufacturing industries and support the development of new industries. Specifically, the President proposes that:
  - NSF increase by $87 million its basic and applied research funding for promising areas of advanced manufacturing technologies, including materials design, nano-manufacturing, next-generation robotics, and cyber-physical systems such as smart buildings.
  - DARPA invest $1 billion in advanced manufacturing over the five years to transform and dramatically decrease the time of

design and production for manufacturing defense systems, vehicles, and other products.

- o NIST laboratories be allocated $760 million to support development of measurements and technological advances in areas including nano-manufacturing, network security, and bio-manufacturing, with increases of $120 million directly for advanced manufacturing.
- o DOE be allocated $500 million to support energy-related advanced manufacturing technologies, such as flexible electronics and ultra-light, ultra durable automotive materials.

- Creation of a public-private partnership program to support platform technologies that will support innovation in manufacturing. The budget calls for $12 million in 2012 to fund these partnerships via the Advanced Manufacturing Technology Consortia (AMTech) program, and for $75 million to be dedicated to PPPs via the Technology Innovation Program (TIP), in particular for technologies that could improve manufacturing processes.
- Incentives for clean energy manufacturing innovation and investment, including the expansion of the Clean Energy Manufacturing Tax Credit from $2.3 billion to $5 billion, with the goal of incentivizing $11.7 billion in private sector investment in clean energy manufacturing projects, firms, and jobs. The current tax credit program has been oversubscribed, and has had proven successful in leveraging private sector investment and creating jobs.
- Investments to support infrastructure for innovation in manufacturing, including establishing a new $3 billion Wireless Innovation Fund to support applied research and development of new communications technologies. The President has also proposed the creation of a National Infrastructure Bank to support high priority projects, sustained investments in high-speed rail, and air traffic control.
- Strengthening of the R&D tax credit. The President has proposed raising the credit from 14% to 17%, simplifying it for use by a broader swath of businesses, and making the credit permanent.
- Investments in improving STEM Education, including the preparation of 100,000 new STEM teachers over the next decade with strong content knowledge and teaching skills and the creation of ARPA-ED, a mission driven agency for education technology to improve learning and aid teaching.

ACKNOWLEDGMENTS

PCAST wishes to express gratitude to the following individuals who contributed in various ways to the preparation of this report as well as the over 100 individuals who sent in comments received via the OpenPCAST website:

Alan Anderson
Writer

Daniel Armbrust
President and CEO
SEMATECH

Robert Atkinson
President
Information Technology and Innovation
Foundation

Norman Augustine
Retired CEO
Lockheed Martin

Asha Balakrishnan
IDA Science and Technology Policy Institute

Ron Bloom
Assistant to the President for Manufacturing Policy
Executive Office of the President

Bruce Brown
Chief Technology Officer Proctor & Gamble

Sujeet Chand
Senior Vice President and Chief Technology Officer
Rockwell Automation

Jim Davis
Associate Vice Chancellor - IT & CIO
Professor Chemical Engineering, UCLA

Willie A. Deese
Executive Vice President and President
Merck Manufacturing Division

Jonathan S. Dordick
Professor and Director
Center for Biotechnology & Interdisciplinary Studies
Rensselaer Polytechnic Institute

Stephen P.A. Fodor
Founder and Executive Chairman
Affymetrix

Erica R.H. Fuchs
Assistant Professor, Department of Engineering
and Public Policy Carnegie Mellon University

Ken Gabriel
Deputy Director
Defense Advanced Research Program (DARPA)

Patrick Gallagher
Director
National Institute of Standards and Technology

Ralph E. Gomory
Research Professor
Leonard N. Stern School of Business
New York University

Matt Ganz
Vice President and General Manger
Boeing Research and Technology

Chuck Grindstaff
CTO and Executive Vice President,
Products Siemens Product Lifecycle
Management Software, Inc.

Krisztina "Z" Holly
Vice Provost for Innovation
Executive Director
USC Stevens Institute for Innovation
University of Southern California

Tom Kalil
Deputy Director for Policy
Office of Science and Technology Policy

John E. Kelly III
Senior Vice President and
Director of Research IBM

Bhavya Lal
Senior Research Staff
IDA Science and Technology Policy Institute

Joel McComb COO
Synthetic Genomics

Lewis Manring
VP of Technology,
Performance Coatings Dupont

Michael J. Natan
CEO
Oxonica Materials Inc

Kelly M. Ogilvie
President and CEO
Blue Marble Energy

Mark Pinto
CTO and Executive
Vice President of Energy and Environmental
Solutions and Display Applied Materials

Dan Reed
Corporate Vice President,
Extreme Computing Program
Microsoft

Leon Sandler
Executive Director,
Deshpande Center for Technological Innovation
MIT

Anna Schorr
Environmental Engineer Nucor Corporation

Stephanie Shipp
Senior Research Staff
Science and Technology Policy Institute

David Stafford
Chief Operating Officer
Michelin Americas Research Company

Marc G. Stanley
Special Advisor to the Director
National Institute of Standards and Technology

Gregory C. Tassey
Senior Economist
National Institute of Standards and Technology

Richard Van Atta
Senior Research Staff
IDA Science and Technology Policy Institute

Gregg A. Zank
CTO and Senior Vice President
Executive Director of Science and
Technology Dow Corning Corporation

End Notes

[1] The balance of trade for "advanced technology products" has widened since 2002, even with a 25% decline in the dollar relative to an index of major foreign currencies. See Gregory Tassey, "Rationales and Mechanisms for Revitalizing U.S. Manufacturing R&D Strategies," *Journal of Technology Transfer 35* (2010): 283-333.

[2] Census Bureau, Foreign Trade Statistics. http://www.census.gov/foreign-trade. Data cited by Tassey, G in "Rationale and Mechanisms for Revitalizing U.S. Manufacturing R&D Strategies", December 2009, Figure 1. http://citeseerx.ist.psu.edu/viewdoc /download?doi=10.1.1.170.3189&rep=rep1&type=pdf

[3] World Industry Service, IHS Global Insight, Inc. Updated: February 17, 2011. This is an estimate of 2010 manufacturing totals that may or may not be confirmed later this year.

[4] Bureau of Economic Analysis, GDP-by-Industry-Accounts, Survey of Current Business 2006-2009, January 2011. Available at: http://www.bea.gov/scb/pdf/2011/01 January/0111_indy_accts_tables.pdf.

[5] International Trade Administration, TradeStats Express, National Trade Data at http://tse.export.gov/TSE/.

[6] Bureau of Labor Statistics, Establishment Data, Historical Employment Data, Table B-1. Employees on nonfarm payrolls by major industry sector, 1961 to date: ftp://ftp.bls.gov/pub /suppl/empsit.ceseeb1.txt

[7] Joel Popkin and Kathryn Kobe, "Manufacturing Resurgence: A Must for U.S. Prosperity," Washington, DC: National Association of Manufacturers and the NAM Council of Manufacturing Associations, January 2010.

[8] Ibid.

[9] Francisco Moris and Nirmala Kannankurry, "New Employment Statistics from the 2008 Business R&D and Innovation Survey," *InfoBrief*, Washington, DC: National Science Foundation, July 2010. (Includes U.S. and abroad employment)

[10] Department of Defense, *Quadrennial Defense Review Report*, February 2010, p. 103.

[11] Bureau of Economic Analysis, GDP-by-Industry-Accounts, Survey of Current Business 2006-2009, January 2011. Available at: http://www.bea.gov/scb/pdf/2011/01January /0111_indy_accts_tables.pdf and http://www.bea.gov/industry

[12] Bureau of Labor Statistics, Current Employment Statistics (National), 2010 ftp://ftp.bls.gov /pub/suppl/empsit.ceseeb1.txt, Table B-1. Employees on nonfarm payrolls by major industry sector, 1961 to date: ftp://ftp.bls.gov/pub/suppl/empsit.ceseeb1.txt

[13] Ron Hira, "The Globalization of Research, Development and Innovation," in *Manufacturing A Better Future for America*, Ed. Richard McCormack, The Alliance of American Manufacturing, 2009.

[14] (In real, inflation-adjusted dollars.) Pisano, Gary P., and Willy C. Shih. "Restoring American Competitiveness." *Harvard Business Review* 87, nos. 7-8 (July - August 2009).

[15] Organisation for Economic Co-operation and Development, Main Economic Indicators, International Trade: Exports in goods (value), online at: http://stats.oecd.org/Index.aspx?DataSetCode=MEI_TRD (last accessed Feb 22, 2011). OECD: http://www.oecd-ilibrary.org/finance-and-investment/ . Export commodities are from the CIA World Fact Book: They provide a listing of the highest-valued exported products, online at. https://www.cia.gov/library/publications/the-world-factbook/fields/2049.html?countryName=Germany&countryCode=gm&regionCode=eu&#gm

[16] Deloitte and U.S. Council on Competitiveness - 2010 Global Manufacturing Competitiveness Index, June 2010. Accessible on the web here: www.deloitte.com/globalcompetitiveness

The report projects that the U.S. will slip from 4th to 5th place in its ranking for manufacturing competitiveness.

[17] Bureau of Labor Statistics, "Production Workers: Hourly compensation costs in U.S. dollars in manufacturing, 34 countries or areas and selected economic groups, 1973–2007," March 2009. (ftp://ftp.bls.gov/pub/special.requests/ForeignLabor/ichccpwsuppt02.txt.

[18] National Science Board's *Science and Engineering Indicators 2010.* Appendix Table 6-19, http://www.nsf.gov/statistics last accessed 2/22/2011

[19] The balance of trade for "advanced technology products" has widened since 2002, even with a 25% decline in the dollar relative to an index of major foreign currencies. See Gregory Tassey, "Rationales and Mechanisms for Revitalizing U.S. Manufacturing R&D Strategies," *Journal of Technology Transfer* 35 (2010): 283-333.

[20] Census Bureau, Foreign Trade Statistics. http://www.census.gov/foreign-trade

[21] Pisano, Gary P., and Willy C. Shih. "Restoring American Competitiveness." *Harvard Business Review* 87, nos. 7-8 (July - August 2009).

[22] National Science Board. *Science and Engineering Indicators 2010* Washington, DC: National Science Foundation. http://www.nsf.gov/statistics

[23] Pisano and Shih, op. cit., p. 3

[24] Ibid.

[25] Note the details of "Why Amazon's Kindle 2 can't be made in the U.S." in Pisano and Shih, op. cit., p. 3

[26] List from Tassey, op. cit., p. i.

[27] National Science Foundation's *Science and Engineering Indicators* 2006 and 2008 and *Research & Development in Industry 2007.* Between 1999 and 2007, foreign R&D funded by U.S. manufacturing firms grew 191% and their funded R&D performed domestically grew 67%, Tassey, op. cit., p. i.

[28] Information Technology and Innovation Foundation. *The Good, the Bad, and the Ugly of Innovation Policy: A Policymaker's Guide to Crafting Effective Innovation Policy.* October 2010.

[29] Even with private investment, China led the United States. See www.bloomberg.com/news /2011-04-03/chinaburies-obama-sputnik-aim-for-clean-power-as-kissinger-advises-bank.html

[30] "Utility-model patents" typically cover items like engineering features in a product and are less ambitious than "invention patents." In the American system, there are no utility patents.

[31] From the "National Patent Development Strategy (2011-2020)" published in November 2009 by the State Intellectual Property Office of China, accessed at http://graphics 8.nytimes.com/packages/pdf/business/SIPONatPatentDevStrategy.pdf, and cited by Steve Lohr, "When innovation too, is made in China", *New York Times,* January 1, 2011

[32] National Research Council. *Innovation Policies for the 21st Century: Report of a Symposium.* 2007.

[33] Fraunhofer Institute for Manufacturing Technology and Advanced Materials IFAM, *Annual Report 2009/2010.* http://www.ifam.fraunhofer.de/jahresberichte/jb09/jb2009_en.pdf

[34] President's Council of Advisors on Science and Technology. *Report to the President and Congress on the Third Assessment of the National Nanotechnology Initiative,* March 2010. Available at: http://www.whitehouse.gov/sites/default/files/microsites/ostp/pcast-nano-report.pdf

[35] See: archive.itif.org/index.php?id=226

[36] From *President's Framework for Revitalizing American Manufacturing:* 40% is due to wages/salaries, 20% due to heath benefits.

[37] Hecker, Daniel. "High Technology Employment: A NAICS-based Update," *Monthly Labor Review* (July 2005): 57- 72. And"High-Technology Employment: A Broader View," *Monthly Labor Review* (June 1999).

[38] Stokes, Bruce. "Act II for American Manufacturing?" *National Journal* (December 2010). http://nationaljournal. com/njonline/no_20100508_1960.php/american-manufacturing-

[39] Lohr, Steve. "G.E. Goes With What It Knows: Making Stutt." *New York Times,* p. BU1 (December 5, 2010). http://www.nytimes.com/2010/12/05/business/05ge.html?src=busln

[40] Aeppel, Timothy. "Coming Home: Appliance Maker Drops China to Produce in Texas." *The Wall Street Journal,* p. B1 (August 24, 2009). http://online.wsj.com/article_ email/SB 125107636394652753-lMyQjAxMTIwNTIxNjAyNzY2Wj.html

[41] New England Council and Deloitte Consulting LLP. *Re-examining advanced manufacturing in a networked world: Prospects for a Resurgence in New England.* Dec 2009

[42] Anderson Economic Group, LLC. *The University Research Corridor's Support for Advanced Manufacturing in Michigan*. July 2010. Accessible at: urcmich.org/news/100721 manufacturing

[43] The study, however, used a somewhat circular definition of "advanced manufacturing" (i.e., sectors with relatively high-paying jobs, involving the making of high-technology products or developing of processes for future manufacturing, and having productivity growth rates significantly above the U.S. average).

[44] U.S. Census Bureau, Local Employment Dynamics. LEHD State of Michigan County Reports. http://lehd.did.census.gov/cgi-bin/qwitop_naicformbrowse? xstate=mi&xstyle=lehd&xntag =31- 33&bktag=&xdbase=county

[45] Bureau of Economic Analysis, NSF and Wolfe 2009 , as cited by Tassey.

[46] Pisano, Gary P., and Willy C. Shih. "Restoring American Competitiveness." *Harvard Business Review 87*, nos. 7-8 (July - August 2009).

[47] Ralph Gomory, " The Innovation Delusion," *The Huffington Post*, March 01, 2010

[48] Tassey, op. cit., p. i

[49] Fleming, Lee, and K. Frenken. "The Evolution of Inventor Networks in the Silicon Valley and Boston Regions." *Advances in Complex Systems* 10, no. 1 (March 2007).

[50] Pisano, Gary P., and Willy C. Shih. "Restoring American Competitiveness." *Harvard Business Review* 87, nos. 7-8 (July - August 2009).

[51] Committee on Critical Technology Accessibility, *Critical Technology Accessibility,* National Research Council (2006).

[52] Recent events in which China has restricted the supply of rare earth metals illustrate this point, especially if China's goal is to move higher up the manufacturing chain so as to preferentially export advanced manufactured materials (e.g., magnets) made from these rare earth metals. Since such magnets are widely used in U.S. defense systems, strategic denial might become a real possibility, against which the U.S. needs to hedge

[53] Department of Defense, *Report to Congress on Implementation of DoD ManTech Projects Receiving FY03 - 05 Funds*, December 2008.

[54] Science Board Task Force on The Manufacturing Technology Program: *A Key to Affordably Equipping the Future Force,* p. 1, February 2006.

[55] It should be noted however, that the U.S. government does provide financing support (through loans and loan guarantees, etc.) for international projects for export-oriented U.S. firms through the Export-Import (Ex-Im) Bank and the Overseas Private Investment Corporation (OPIC).

[56] OECD Tax Database, http://www.oecd.org:80/ctp/taxdatabase

[57] PCAST notes that corporate tax structures differ across nations, with the result that the actual marginal tax rate for some corporations may differ significantly from the rate in the tax codes; this complicates international comparisons. (See: Kocieniewski, David. "U.S. Business has High Tax Rates but Pays Less," May 2, 2011. http://www.nytimes.com/2011/05/03/business/economy/03rates.html). We also note that it is important to consider the relative merits of lowering marginal tax rates versus providing investment-specific incentives.

[58] The "R&D tax credit" is used in this report to refer to the U.S. Research and Experimentation Tax Credit.

[59] Deloitte and U.S. Council on Competitiveness - 2010 Global Manufacturing Competitiveness Index, June 2010. Accessible on the web here: www.deloitte.com/globalcompetitiveness

[60] Tassey, op. cit., p. i.

[61] Block, Fred and Keller, Matthew. "Where do Innovations Come From? Transformations in the U.S. National Innovation System, 1970-2006" *Information Technology and Innovation Foundation Report.* July 2008. As recognized by *R&D Magazine.*

[62] The current state of the technology consists of silicon photonics that integrate photonic and electronic circuits, monolithically integrated into single substrates with bonded lasers fabricated from indium phosphide. The next generation of this technology will require substantial investment in developing processes that are 1) compatible with existing CMOS fabrication facilities, 2) stable and repeatable, and 3) accessible to the wider community and 4) offer bestin-class performance for both photonic and electronic devices.

[63] We note that PCAST has not sought to evaluate how the overall funding portfolio might be rebalanced. Instead, we have identified gaps and needs that are not being met currently.

[64] These might include specific technology areas such as the examples in Box 2, or the development of new production and measurement equipment.

[65] National Academy of Sciences, National Academy of Engineering, Institute of Medicine, *Rising Above the Gathering Storm: Energizing and Employing America for a Brighter Economic Future.* Washington, DC: National Academies Press. 2005.

[66] Federal R&D was equivalent to 1.92% of GDP in 1964, and 0.76% of GDP in 2004. http://www.nsf.gov/statistics

[67] For example, in the Programme for International Student Assessment (PISA), which measures students'ability to apply what they have learned in science and technology and has been designed to assess the kinds of skills needed in today's workplace, U.S. 15-year-olds scored below most other nations tested in 2006. The U.S. standing dropped from 2000 to 2006 in both mathematics and science. Also, the World Economic Forum ranks the United States 48th in quality of mathematics and science education.55 World Economic Forum, The Global Competitiveness Report 2009-2010, Available at: http://www.weforum.org /pdf/GCR09/Report/Countries/United%20States.pdf.

[68] President's Council of Advisers on Science and Technology. *Prepare and Inspire: K-12 Education in Science, Technology, Engineering and Mathematics for America's Future.* http://www.whitehouse.gov/sites/default/files/microsites/ostp/pcast-stemed-report.pdf

[69] www.whitehouse.gov/innovation/strategy

[70] Currently, more than half of the 140,000 employment-based visas are used to admit spouse and children.

In: United States Manufacturing
Editors: V. Murphy and J. E. Walker ISBN: 978-1-62081-522-9

Chapter 5

DEPARTMENT OF COMMERCE: OFFICE OF MANUFACTURING AND SERVICES COULD BETTER MEASURE AND COMMUNICATE ITS CONTRIBUTIONS TO TRADE POLICY*

United States Government Accountability Office

WHY GAO DID THIS STUDY

Declining U.S. manufacturing has been an issue of continuing concern for policymakers; this was reflected in the Obama Administration's (Administration) 2010 announcement of the National Export Initiative. The Administration has also shown interest in improving the efficiency of the federal support of trade operations. In 2004, the Office of Manufacturing and Services (MAS) was established within the Department of Commerce's (Commerce) International Trade Administration (ITA) to enhance the global competitiveness of U.S. industry.

GAO was asked to examine (1) MAS's goals and activities and how they compare with those of other government entities; (2) how MAS prioritizes its activities and targets its resources; and (3) the extent to which MAS tracks and

* This is an edited, reformatted and augmented version of United States Government Accountability Office, Report to Congressional Requesters, Publication No. GAO-11-583, dated June 2011.

reports its efforts. GAO reviewed agency documents and interviewed officials from MAS, other parts of ITA and Commerce, and other agencies.

What GAO Recommends

GAO recommends that the Secretary of Commerce take actions, in concert with MAS, to finalize MAS's focusing of mission and priorities, systematically monitor workload, and more systematically obtain and communicate information on the value MAS adds to the trade policy process. In its comments, Commerce concurred with the findings and recommendations and expects to make progress by October 2011.

What GAO Found

MAS's primary goal is to support the competitiveness of U.S. industry, which it does largely through combining its industry and trade expertise to support other parts of Commerce, including other parts of the ITA (see figure below) and external U.S. government clients, such as the Office of the U.S. Trade Representative (USTR). The major activities of MAS's offices include: collection and dissemination of data on U.S. industry and trade, production of analyses on policies that can affect competitiveness, and identification and resolution of overseas trade barriers. While some activities may seem similar to those of other agencies, such as USTR, officials from MAS's client agencies stated that MAS's combination of industry and trade expertise is not readily available to them elsewhere in the government.

MAS has undertaken an internal review to update its mission and priorities regarding activities and clients and has proposed changes currently under departmental review. MAS does not have a mechanism to systematically monitor analysts' workload or the amount of time spent on requests for different clients. The absence of workload data may hinder its ability to effectively allocate its resources to address the needs of the trade policy process. Further, MAS's role has not been clearly communicated, and ITA's Web site provides limited information about MAS. Consequently, the public and Congress have limited information about MAS's activities and contributions to policy making.

MAS's ability to meet its performance targets largely depends on actions from other government agencies and other parties, making isolating its contributions difficult. MAS developed a series of steps, or milestones, to help isolate its contributions to trade policy outcomes, although officials acknowledged continuing challenges. Further, MAS does not systematically obtain feedback on its performance from the agencies to which it provides analysis, nor does it track its contributions to major policy decisions that fall outside its externally reported performance targets. This makes it difficult to assess the extent to which MAS's work adds value to the trade policy process.

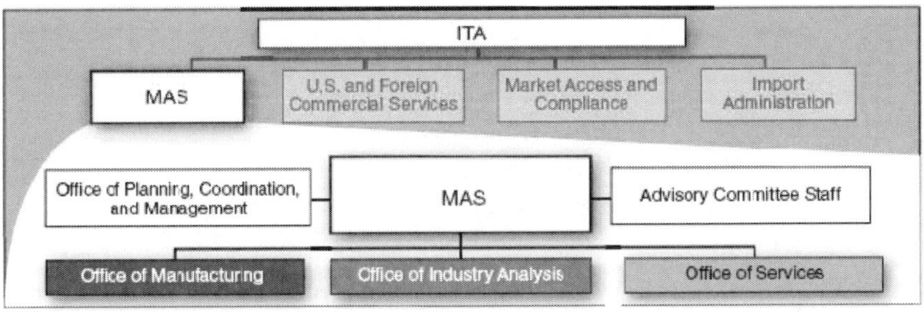

Source: GAO analysis of ITA dataurce data.

Office of Manufacturing and Services in the ITA.

ABBREVIATIONS

Administration	Obama Administration
APEC	Asia-Pacific Economic Cooperation
BEA	Bureau of Economic Analysis
Board	ITA's Web Governance Board
Commerce	Department of Commerce
Committee	Committee on Foreign Investment in the United States
DFO	designated federal officers
EU	European Union
GDP	Gross Domestic Product
IT	Information Technology
ITA	International Trade Administration
MAS	Office of Manufacturing and Services
MDCP	Market Development Cooperator Program

MRA	Mutual Recognition Agreement
OMB	Office of Management and Budget
PART	Program Assessment Rating Tool
State	Department of State
TPCC	Trade Promotion Coordination Cabinet
TPIS	Trade Policy Information System
USITC	U.S. International Trade Commission
USTR	Office of the U.S. Trade Representative

June 7, 2011
The Honorable Sherrod Brown
Chairman
Subcommittee on Financial Institutions and
Consumer Protection Committee on Banking,
Housing, and Urban Development
United States Senate

The Honorable Jeff Merkley
United States Senate

Declining U.S. manufacturing, particularly manufacturing jobs, has been an issue of continuing concern for policymakers. This concern was reflected in the Obama Administration's (Administration) 2010 announcement of the National Export Initiative, which brought new emphasis to one proposed government response to manufacturing declines—a renewed federal focus on increasing U.S. exports.[1] In addition, the Administration has demonstrated increased interest in improving the efficiency of the multiple federal agencies that support trade operations, with President Obama requesting that the Office of Management and Budget (OMB) prepare a reorganization proposal by June 2011. Recently, Congress expressed interest in the activities and achievements of the Office of Manufacturing and Services (MAS) within the Department of Commerce (Commerce), which was established to enhance the global competitiveness of U.S. industry, expand market access, and increase exports.

In response to your request, we examined: (1) the goals and activities of MAS and how the types of analysis and expertise MAS provides compare with those provided by other government entities; (2) how MAS prioritizes its activities and targets its resources; and (3) the extent to which MAS tracks and reports its contributions to increasing U.S. competitiveness and trade. In addition, you asked us to provide information on U.S. and global manufacturing trends (see app. II).

To address these objectives, we reviewed relevant documents, including legislative authority and budget and staffing information; documents related to MAS's performance measurement system; and examples of how MAS's work has been used by parts of Commerce and other U.S. government agencies over the last 5 years. In addition, we discussed MAS's efforts with its officials, its government clients, and others who are knowledgeable about MAS. We interviewed Commerce officials from the Offices of the Secretary, Commercial Service, Market Access and Compliance, Import Administration, and the National Institute of Standards and Technology. We also interviewed officials from other U.S. government agencies, including the U.S. International Trade Commission (USITC), the Office of the United States Trade Representative (USTR), Departments of State (State), Energy, Transportation, and Treasury, the Environmental Protection Agency, and OMB. Appendix I provides more information on our scope and methodology.

We conducted this performance audit from August 2010 to June 2011 in accordance with generally accepted government auditing standards. Those standards require that we plan and perform the audit to obtain sufficient, appropriate evidence to provide a reasonable basis for our findings and conclusions based on our audit objectives. We believe that the evidence obtained provides a reasonable basis for our findings and conclusions based on our audit objectives.

BACKGROUND

MAS is a 207-person unit within Commerce's International Trade Administration (ITA), as shown in table 1. ITA's stated mission is to strengthen the competitiveness of U.S. industry by promoting trade and investment and by monitoring and enforcing U.S. trade laws and agreements.[2] MAS is one of four distinct, but interrelated, business units within ITA, each led by an Assistant Secretary.

MAS was created in 2004, after the publication of Commerce's *Manufacturing in America* report which called for its creation to support the Secretary of Commerce in his role as the federal government's chief advocate for the manufacturing sector. The report grew out of a 2003 review of the U.S. manufacturing sector initiated by former Secretary of Commerce Donald Evans in response to "unprecedented challenges" facing U.S. manufacturers.

Commerce received input from industry associations and large and small manufacturers from critical sectors, and the report summarized manufacturers' concerns, which included the government's limited focus on manufacturing and its ability to compete globally. In addition to other recommendations, the report called for the creation of an Assistant Secretary of Commerce for Manufacturing and Services.

Table 1. International Trade Administration Offices—Budget, Staffing, and Activities

Dollars in millions			
ITA business unit	Fiscal year 2010 budget	Full-time equivalents	Activities summary
Manufacturing and Services	$49.5	207	MAS industry experts and economists perform research and analysis for internal government clients on trade policy and competitiveness issues. MAS also provides support to the private sector when there is a wider industry focus.
U.S. and Foreign Commercial Service	260.4	1,041	Commercial Service promotes and protects U.S. commercial interests abroad and advises individual businesses to enhance firms' global competitiveness.
Market Access and Compliance	44.7	204	Market Access and Compliance uses its regional expertise to assist U.S. companies facing trade barriers in foreign markets by working to remove those barriers and contributes regional and country expertise to trade policy making.
Import Administration	69.1	331	Import Administration is the lead unit on enforcing trade laws and agreements to prevent unfairly traded imports and to safeguard jobs and the competitive strength of U.S. industry.

Source: GAO analysis of ITA information and data.

Congress made recommendations in a 2003 House of Representatives Appropriations Committee Report to realign ITA's structure and clarify the mission of each business unit to better address an increasingly competitive global economy and growing U.S. trade deficit.[3] The report called for the creation of a better analytic basis for U.S. trade policies and negotiations. It contained several specific actions, including ones targeting U.S.

manufacturers. For example, Congress expected the proposed MAS unit to develop tools and expertise to assess industry trends and evaluate the impact of trade agreements, and to identify and address challenges facing manufacturers through an interdepartmental advisory committee. Subsequent legislative and agency activity established the reorganization proposed in the committee report and transferred functions and staff across ITA business units, with one result being the reorganization of the Trade Development unit into MAS.[4]

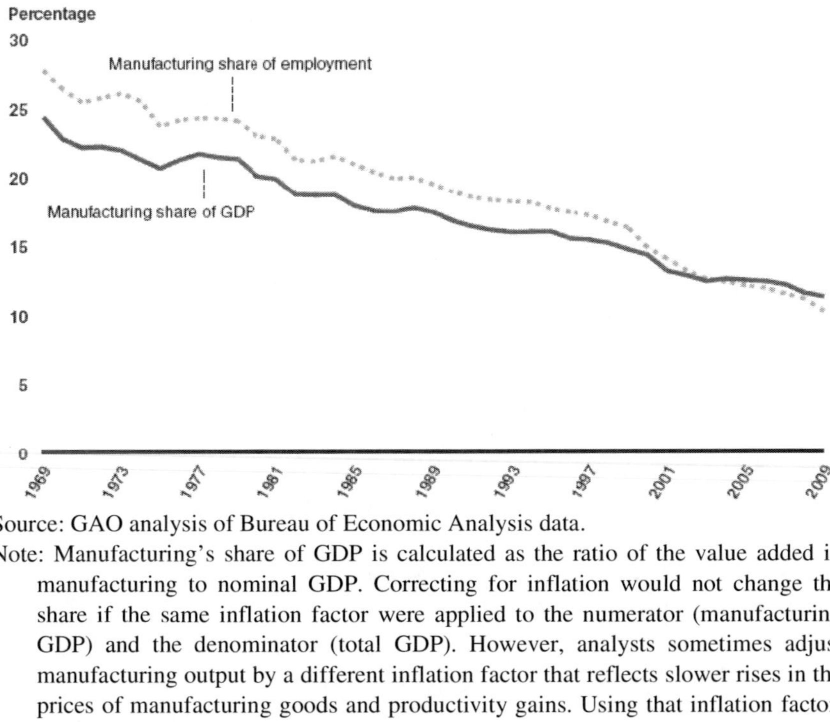

Source: GAO analysis of Bureau of Economic Analysis data.

Note: Manufacturing's share of GDP is calculated as the ratio of the value added in manufacturing to nominal GDP. Correcting for inflation would not change the share if the same inflation factor were applied to the numerator (manufacturing GDP) and the denominator (total GDP). However, analysts sometimes adjust manufacturing output by a different inflation factor that reflects slower rises in the prices of manufacturing goods and productivity gains. Using that inflation factor, manufacturing's share of real output would have remained roughly constant over this period.

Figure 1. GDP and Employment in the Manufacturing Sector Relative to the Overall U.S. Economy, 1969 – 2009.

Congressional concerns leading to the reorganization of ITA stemmed, to a large degree, from continuing trends in manufacturing and trade. Although the United States remains the world's largest producer of manufactured products, there have been steep declines in manufacturing employment as measured by share of hours worked, and manufacturing's share of gross domestic product (GDP) (see fig.1). Manufacturing employment fell from about 28 percent of total U.S. employment in 1969 to about 10 percent in 2009. As a share of nominal GDP, the drop has been similar. However, even with declines, almost 10 percent of the U.S. economy is in manufacturing—roughly 11 million workers. In addition, manufacturing continues to account for a large share of U.S trade. In 2009, manufactured goods accounted for about 60 percent of U.S. exports.

The National Export Initiative brought new emphasis to the federal government's role in promoting exports.[5] Commerce, with ITA as the lead entity within Commerce, works with other federal government agencies on the Export Promotion Cabinet, which, in collaboration with the Trade Promotion Coordinating Committee (TPCC), is charged with carrying out the initiative and spurring job growth through the doubling of U.S. exports by 2015. ITA also works through the USTR-led interagency structure used to formulate trade policy and represent U.S. trade interests in multilateral and bilateral forums such as the World Trade Organization. MAS analysts cover essentially all nonagricultural sectors of the economy.

MAS Combines Industry and Trade Expertise in Providing Analytical and Policy Support to the U.S. Government

MAS Offices Focus on Industry Sectors and Economic Analysis, Primarily in Manufacturing

MAS's primary goal is to support the competitiveness of U.S. industry in domestic and international markets, which it does largely through providing policy advice, research, and analytical support to other parts of Commerce and the U.S. government.[6] MAS has distributed 182 of its 207 staff across three suboffices, with the largest share in its Office of Manufacturing. As shown in figure 2, analysts in the Office of Manufacturing and the Office of Services cover a number of industry sectors, serving as sources of industry information from a trade perspective. Analysts and economists in the Office of Industry Analysis conduct economic and policy analysis to support U.S. industry and evaluate industry recommendations for trade negotiations and U.S. competitiveness.

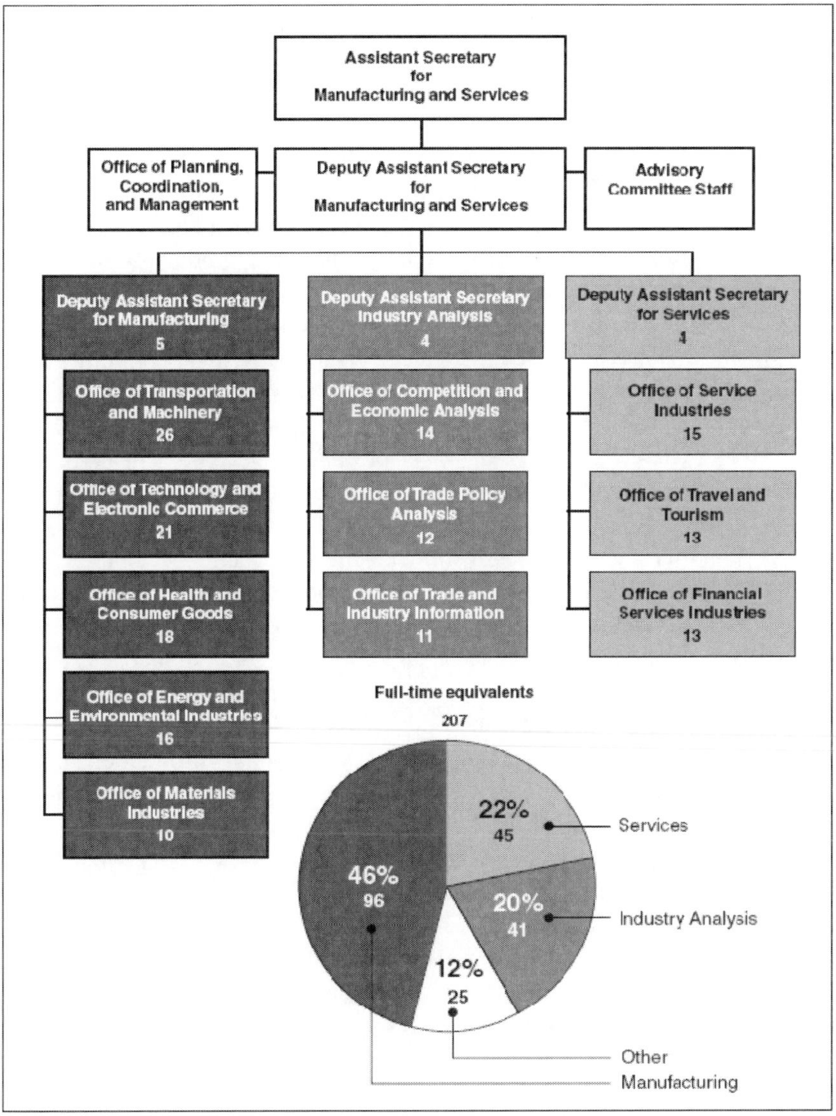

Source: GAO analysis of MAS data.

Figure 2. Distribution of Staff (Full-Time Equivalents) Across MAS Offices, as of January 2011.

MAS Supports Trade Policy Efforts of Internal and External Government Clients Mainly through Industry Research and Sector-Specific Analysis

MAS analysts provide industry research and sector-specific analysis to support trade policy efforts of both internal clients in Commerce and external government clients, such as USTR. MAS's major activities include: collection and dissemination of data on U.S. industry and trade; production of analyses on domestic and international trade and investment policies that can affect competitiveness; identification and resolution of overseas market and trade barriers; and management of the Industry Trade Advisory Committees (see tables 2 and 3 for more detailed descriptions of MAS's activities; see app. III for examples of industry-specific trade barriers that MAS addressed in fiscal year 2010). The 2004 reorganization of Trade Development into MAS transferred day-to-day servicing of private sector requests to the Commercial Service offices in the field.[7] The portion of MAS's research and analysis that is used primarily by internal government clients (including Commerce and ITA business units, as well as other executive branch agencies such as USTR and the Department of the Treasury) for sensitive negotiations and policy making is not publicly available. However, some of the industry-specific information and data produced by the Office of Industry Analysis is made publicly available. MAS provides other parts of ITA and Commerce with industry sector analyses, which contain background information on the specific industry; evaluation of its competitive strengths and weaknesses both domestically and internationally; and proposed strategic direction for industry sectors (see app. IV for more information on MAS's sector analyses). According to agency officials, other ITA units use these analyses to address specific trade issues that affect industries and regions. Table 2 shows examples of the types of support MAS provides to its internal Commerce clients. MAS also supports the efforts of other executive agencies, including USTR, through activities that include review and analysis of trade agreements, tariffs, and domestic regulations; involvement in the interagency trade policy-making process; and serving as a source of trade data to the government and public. USTR officials noted the contribution of MAS's expertise and analysis to the U.S. trade policy process. According to USTR officials, MAS's position in government, its insight into industry, and its analysis through a trade focused viewpoint are utilized by USTR in the trade negotiating process. A USTR official further stated that, in addition to MAS's role in supporting the trade policy process, it also engages directly in trade promotion activities that benefit U.S. companies. USTR noted that MAS can play an important role in providing insight into industry's concerns and potential reactions to policy changes. USTR officials also commented that MAS provided useful information to them during the U.S.-Korea free trade agreement negotiations. For example, MAS provided detailed trade data on autos traded according to engine sizes, which could affect the application of certain tariffs. According to USTR officials, other government or private sector entities do not generally have the capacity for the specialized trade and tariff data that MAS can produce.

Table 2. Examples of MAS Activities that Support Internal Commerce Clients, Including Other ITA Business Units

MAS Works with Other ITA Business Units to Address Specific Trade Issues that Affect Industries and Regions
• *Market Access and Compliance. MAS provides information about a particular industry as a whole to Market Access and* Compliance analysts who focus on market access issues in specific regions. MAS provides industry-wide information and analysis on broader trade barriers that affect more than one company. A Market Access and Compliance official stated that its Trade Compliance Center involves MAS industry experts on every case.
• *Commercial Service. MAS responds to the needs of Commercial Service officers who are advising companies in the field by* providing industry-specific information. For example, a Commercial Service official stated that she contacted MAS industry analysts who had knowledge of the agricultural machinery sector to obtain information for a U.S. company interested in exporting irrigation equipment to Eastern Europe.
• *Import Administration. MAS provides industry information to Import Administration in its review of overseas companies'* applications for foreign trade zone designation. For example, an Import Administration official said that her office relies on MAS to ensure that its review of applications for special customs treatment is consistent with U.S. trade policy and will not adversely affect other companies.
MAS Provides Senior Commerce Officials with Briefing Materials
• MAS provided export and trade data to the Secretary of Commerce for the President's trip to India.
• MAS worked with Commerce's Economic Development Administration, National Oceanic and Atmospheric Administration, and Minority Business Development Administration to provide the Secretary of Commerce and White House officials with an analysis of the economic impact of the Deepwater Horizon Gulf oil spill and, more specifically, the impact on the Gulf's travel and tourism industry.

Source: GAO analysis of MAS information and data.

USTR officials stated they were unable to provide the actual documents MAS provided to USTR because they were part of the internal deliberative process used to develop U.S. negotiating positions. However, they did state that the type of analyses are not unlike MAS state and sector analyses, which are available to the public.

Table 3. Examples of MAS Activities that Support the Efforts of Other Executive Agencies

MAS Supports USTR's Trade Negotiations by Helping to Develop U.S. Positions on Tariffs, Government Procurement, and Rules of Origin
• MAS recommends the tariff treatment for industrial products in trade negotiations and compiles tariff offers and requests for U.S. trading partners.
• MAS develops recommendations for USTR on U.S. government procurement offers and requests for improvement based on industry input, assessments from MAS industry analysts, and research of the trading partners' government procurement market.
• MAS ensures that U.S. industry positions on product-specific rules of origin are considered during negotiation of trade agreements.
MAS Analysis Supports Government Efforts to Implement and Enforce U.S. Trade Laws
• MAS analyzes product petitions submitted under the Generalized System of Preferences program and provides recommendations for Commerce's positions on these petitions.
• MAS creates lists for the interagency process identifying imports from countries to which the U.S. government potentially will apply additional duties in retaliation for unfair trade practices, pursuant to Section 301 of the U.S. Trade Act of 1974.
• MAS reviews and analyzes the Miscellaneous Tariff and Trade bills and coordinates the interagency process for developing Administration recommendations for consideration by Congress.
MAS Provides Industry's Perspective to the Interagency Policy-making Process
• MAS provides support to the Industry Trade Advisory Committee system, which Commerce jointly administers with USTR. The advisory committees allow representatives from private business and other groups who are cleared advisors, and whose companies engage in trade to provide input with respect to ongoing trade negotiations and trade policy.
• MAS provided guidance on the renegotiation of the OECD Aircraft Sector Understanding, the special set of rules governing support for civil-aircraft exports.
MAS Provides Analysis on Certain Domestic Industry Issues
• MAS analyzes the impact of regulations and legislation on U.S. industry and provides recommendations to regulatory agencies such as the Environmental Protection Agency.
• MAS provides industry analysis to the Department of the Treasury on cases before the Committee on Foreign Investment in the United States, which considers the national-security implications of certain foreign mergers, takeovers, and acquisitions in the United States.
MAS Is a Source for Data and Statistics on Exporters, Trade, and Industry that Are Used by the Government and Private
Sector
• MAS is responsible for the management of the travel and tourism statistical system for assessing the economic contribution of the industry. As part of this work, MAS collects, analyzes, and disseminates international travel and tourism statistics.
• MAS is one source of accessible and comprehensive trade data for government (Trade Policy Information System (TPIS)) and general public (TradeStats Express) users. The trade data that MAS provides includes the Metropolitan Export Series and the Exporter Database, which contain subnational export data.

Source: GAO analysis of MAS information and data.

Table 4. Differences between MAS's Activities and Clients and Those of Other Executive Agencies

MAS Focuses on Industry and Trade Policy, while Other ITA Units Have Regional Expertise and Counsel Individual Businesses
• MAS's counsel of individual businesses is limited to sector-specific policy and promotional issues. In contrast, U.S. and Foreign Commerce Service officers are placed in domestic and foreign field offices to work directly with individual businesses to help them expand their exporting capacity and access foreign markets.
• MAS analysts are organized by industry and sector and work to address industry-wide trade policy barriers, rather than those affecting individual firms. In contrast, Market Access and Compliance analysts work by geographic region and respond to specific market-access issues faced by a particular U.S. firm.
• Import Administration is responsible for enforcing U.S. trade laws and responds to complaints and petitions from individual firms. MAS's involvement in trade enforcement includes providing industry and sectoral analysis as it pertains to resolving certain trade disputes.
Like MAS, USITC Has Industry Offices, but It Acts Principally as a Technical Advisor to the Interagency Policy-making Process and Provides Its Formal Reports and Advice in Response to Official Requests
• USITC is an independent, quasi-judicial federal agency. As such, it serves as a nonvoting member of the Trade Policy Staff Committee.
• USITC prepares reports under Section 332 of the Tariff Act only at the request of the President (who has delegated that authority to USTR), the House of Representatives Committee on Ways and Means, the Senate Finance Committee, the full House of Representatives, or full Senate.
• USITC provides information and analysis, but, unlike MAS, does not provide policy recommendations. USITC reports are made available to the public, except those that are classified as national security information by USTR. USITC redacts confidential business information from all public reports.
MAS's Trade Specific Industry Expertise Differs from the Technical Expertise Found in other Government Agencies
• The Department of Energy employs engineers and scientists with specific expertise in the relevant technologies utilized in the energy sectors, not expertise in trade or the industry as a whole. Department of Energy officials noted that in contrast, MAS analysts typically have international business backgrounds, not scientific ones.
• Officials from the Department of Transportation stated that MAS's knowledge of trade policy complements their technical understanding of engineering and infrastructure, allowing them to better serve industry and move their issues to a more prominent position in the National Export Initiative.

Source: GAO analysis of Commerce and other agency information.

That analysis is developed in collaboration with other ITA units for each free trade agreement and are posted on USTR's Web site, as well as ITA's. These analyses highlight the new market opportunities that a particular trade agreement provides to U.S. exporters and the effects of international trade on all 50 states' economies.

In addition to supporting USTR, MAS plays a key role within Commerce in providing support to other government agencies in areas such as supporting trade negotiations and providing data on travel and tourism, a leading U.S. services export. According to MAS officials, MAS provides analysis of and policy advice to Commerce on transactions seeking official financing from U.S. government agencies and multilateral development banks in which the U.S. government has a vote, as well as on key official finance issues under discussion in multilateral forums. These include, for example, the U.S. Export-Import Bank, Overseas Private Investment Corporation, and the Organisation of Economic Cooperation and Development Exports Credits Group. Table 3 provides specific examples of the types of MAS activities that support other executive branch agencies.

MAS's Clients Report that MAS Provides Analysis Not Readily Available Elsewhere in Government

While MAS conducts activities that have similarities to activities of other agencies, officials from MAS's client agencies stated that MAS can provide analysis that combines industry and trade expertise that is not readily available elsewhere in government. For example, government officials we interviewed stated that some other agencies may have technical expertise in particular disciplines but have less of a focus on developing trade policy. Table 4 provides examples of how MAS differs from other ITA units, the U.S. International Trade Commission, and other government agencies.

MAS Faces Challenges in Prioritizing Activities, Monitoring its Workload, and Communicating its Role and Contributions

MAS Has Undertaken an Internal Review, but Has Not Had a Clear System to Prioritize Activities or a Clear Mission Statement

MAS has undertaken an internal review to update its mission and priorities regarding activities and clients and has proposed changes which are currently under departmental review. Since the 2004 reorganization of ITA,

MAS's broad mission statement has not clearly defined its role in government. MAS's stated mission has been to strengthen the competitiveness of U.S. manufacturing and services by addressing commercial and economic impediments that disadvantage U.S. companies overseas. However, ensuring the competitiveness of U.S. industry requires policies and actions from many government agencies, not just Commerce. For example, U.S. corporate tax policy affects the competitiveness of U.S. industry, but Commerce does not have decision-making authority to regulate tax policy.

MAS officials have acknowledged the need for having clearer priorities and mission alignment with the Administration's National Export Initiative's goals to successfully serve its clients. A MAS official stated that MAS cannot be "all things to all people," which has been a consistent concern raised by staff. To address this need, in early summer 2010, MAS officials began an internal review of its mission and activities. As a result, MAS revised its mission statement: to advance the competitiveness of U.S. industries by leveraging its in-depth sector expertise in the development of trade policy and promotion strategies. MAS officials stated that the changes made as a result of the internal review are still in process and must be approved at the departmental level.

MAS has an annual planning process to identify industry issues and determine actions to address them, but MAS officials acknowledged that these plans do not capture the numerous day-to-day unanticipated requests that come from sources such as the Secretary of Commerce, Executive Office of the President, other Executive Branch agencies, and the Congress. Every year, each industry office develops an assessment of the industry sectors it covers which, MAS guidance states, should present an analysis of the competitive strengths and weaknesses of the industry and an assessment of the industry's needs. Each office then uses these industry assessments to develop business plans, which describe activities the office plans to undertake during the year, and links business plan activities to objectives, performance measures, and targets. For example, the fiscal year 2010 business plan for MAS's Office of Energy and Environment included identifying market access opportunities and barriers, as part of its energy-efficiency initiative. The plan linked this activity to MAS's performance measure to identify the percentage of industry-specific trade barriers that were removed or prevented. MAS officials told us that its management reviews the plans and uses them in making decisions throughout the year. However, MAS managers told us they prioritize the unanticipated requests based on resource availability and importance of the activity.

As part of its internal review, MAS has developed decision criteria to assist analysts in prioritizing their work demands. According to the proposed criteria, which is in process awaiting departmental approval, MAS will primarily support sectors that have a direct connection to exports or that strategically impact trade. Such a sector must be a high-volume exporter, with $10 billion or more in exports or be a high potential-growth sector such as renewable energy. MAS management has set the goal of having 75 percent of MAS's resources working toward National Export Initiative objectives of doubling exports by 2015. However, MAS officials noted that they currently conduct work that is important for U.S. global competitiveness but not directly related to exports. For example, MAS provides the business perspective on cases that go before the Committee on Foreign Investment in the United States (Committee), and coordinates Commerce participation in these cases. The Committee is an interagency panel authorized to review transactions that could result in control of a U.S. business by a foreign person. The review is to determine the effect of such transactions on the national security of the United States and is not directly related to exports. To align with its new approach for prioritizing analysis related to exports, MAS management plans to streamline resources devoted to working on those cases that are not of strategic importance. In addition, MAS plans to reduce its effort in analyzing domestic regulations that do not have a significant impact on exports.

MAS's System to Monitor Its Workload is Used Unevenly

MAS does not have a mechanism to systematically monitor analysts' workload or the amount of time spent on requests for different clients. MAS officials identified several top-priority clients, but the many unanticipated daily requests can pose difficulties for MAS analysts. Some requests for the Offices of Services and Manufacturing come to analysts through ITA's formal "tasker" system and are delegated by senior management to analysts. The "tasker" system is an electronic system through which tasks are assigned to staff and signed off on by management. MAS officials stated that it is useful for tasking out assignments to multiple offices; however, they noted that it does not cover all the work undertaken within MAS. They noted that for the Office of Industry Analysis, most requests are directly communicated to managers and analysts by a colleague in ITA or another government agency. They stated that entering Office of Industry Analysis workload data into the tasker system is not current practice. Moreover, according to MAS officials, many of the requests involve sensitive documents which cannot be entered into the system in order to comply with security protocols. MAS managers said that the proportion of work captured by the tasker system varies greatly among its program areas, ranging from 85-90 percent for the Offices of Manufacturing and Services to less than 5 percent for the Office of Industry Analysis. Without a way to systematically monitor staff's workload, it is difficult for management to determine how to most efficiently allocate resources.

In prior work, we found that for an entity to run and control its operations, management must have relevant, reliable, and timely information.[8] However, MAS did not systematically track the time spent on tasks by different types of clients. After our inquiries, the Deputy Assistant Secretary of MAS's Office of Industry Analysis began monitoring short-term (2 to 3 days) requests by different types of clients and found that approximately 70-75 percent of short-term requests for analysis came from within ITA, 5-10 percent were generated from the rest of Commerce, and the remaining 15-20 percent were from other agencies in the Executive Branch and the Congress. However, these data reflect only the requests— which average 5,400 annually, according to MAS officials—submitted to the 41 staff in Office of Industry Analysis, not MAS offices overall, and they do not include information that involves security protocols or longer-term projects. According to agency officials, these short-term requests are ad hoc and often time sensitive and must be balanced with the broader long-term activities MAS analysts undertake. Given that management does not have reliable and timely information about staff's workload, it may be difficult for management to ensure staff are working on the highest-priority efforts.

MAS Faces Challenges in Communicating Its Contributions to Congress and the General Public

Officials from agencies that work with MAS told us they understand MAS's role and contributions to trade policy and competitiveness issues, but these contributions are not readily apparent to Congress or the general public. This is partly because MAS's policy-development work is not publicly available. Although ITA's Web site does provide public access to industry, trade, and economic data and analysis produced in MAS, it provides limited information about MAS's role and activities. Consequently, the public and Congress may have limited information about MAS's contributions to policy making. This lack of transparency may hinder Congress's ability to provide effective oversight of MAS's activities.

There is little publicly available information about MAS's priorities and contributions to policy making. As noted above, MAS's work is largely intended for use by internal government clients, and much of that output is not in the form of products that are externally available. Previously, we found that, in addition to adequate internal communications, management should ensure there are adequate means of communicating with external stakeholders, which can include Congress.[9]

An organization's Web site is typically a readily accessible source of information to the public about the organization's mission and activities. A MAS official stated that when she worked in the private sector, she relied on the Web site to get the analysis and information they needed and that its Web sites were organized in a manner useful for industry users. In addition, officials from agencies that rely on MAS for analytical support stated they generally communicate with MAS via phone or e-mail and do not rely on the Web site. Further, MAS and other officials stated that ITA's internal Web site, or intranet, provides a primary means of communication among ITA units and, thus, a public ITA Web site is less important for that purpose.

For the general public, ITA's Web sites provide limited information about MAS's priorities and activities. According to a 2007 Commerce Inspector General report, ITA's Web sites have duplicative, confusing, disorganized, and outdated pages.[10] In a December 2010 meeting, members of ITA's Web Governance Board (Board) told us that the issues raised in the report are still relevant. The board observed that Web sites are important tools to increase public awareness of ITA programs and expertise, not only for the business community, but also for external stakeholders who may have difficulty finding useful information about MAS as an organization and its role in trade policy.

The board members told us that because ITA does not have a centrally directed web management office, each business unit is responsible for managing its own web presence, but there is great variation among the offices' ability and technical expertise. A MAS official stated that, MAS, like some of the other ITA business units, does not have dedicated full-time staff focused on maintaining MAS's web presence. Instead there are several MAS analysts who have the auxiliary duty of working on MAS's Web site in addition to their other duties. Consequently, MAS and ITA as a whole lack cohesive, transparent, and consistent Web sites. This may also hinder MAS's ability to effectively communicate its priorities and contributions to stakeholders, Congress, and the public.

MAS Faces Challenges in Measuring Its Contributions to the Trade Policy Process

External Factors Affect MAS's Achievement of Its Broad, Outcome-Based Performance Targets

Due to its policy-support role, MAS's ability to meet its performance targets, such as breaking down trade barriers faced by U.S. firms, depends on actions from other agencies, Congress, businesses, and foreign governments. MAS has three main objectives linked to five outcome-based performance measures[11] that it reports on externally in assessing its performance.[12] One objective—to ensure appropriate industry and other stakeholder input into trade policy development, negotiations, and implementation—covers much of what MAS does, including addressing industry-specific trade barriers and working with USTR on trade agreements. MAS has also developed internal measures that are used to track activities and for planning, but are not reported externally (see tables 5 and 6).

MAS's fiscal year 2011 budget justification stated that one limitation to its meeting its performance targets is that many factors—including U.S. business cooperation, global trade trends, political developments, and the extent to which foreign governments create barriers or act inconsistently with trade obligations—affect the number of barriers removed.

MAS's performance-measurement process attempts to address this challenge by breaking its outcome-based measures into specific milestones, according to officials.

The specific milestones—such as identifying market-access issues of interest to U.S. industry or industry-specific opportunities resulting from a trade agreement—are analyses or industry-outreach steps that MAS itself can undertake and complete, even though MAS cannot control the ultimate impact of those actions. MAS sets targets for its performance measures annually. These milestones, while more specific than the overall performance measures, are nonetheless qualitative measures.

In Commerce's fiscal year 2010 *Performance and Accountability* report, MAS reported that it had met or exceeded all of its targets for its four performance measures.[13] For example, MAS reported that it had met its target of removing or preventing 30 percent of the industry-specific trade barriers it addressed.

In both 2008 and 2009, Commerce's *Performance and Accountability* reports stated that MAS exceeded three of its targets and did not meet its target for its work on trade agreements, because the Administration suspended work on two trade agreements MAS was working on.

Table 5. MAS Objectives and External Performance Measures Used to Enhance U.S. Competitiveness in International Markets

Objective	Performance measure (external—for Government Performance and Results Act)	Examples
Ensure industry and other stakeholder input into trade policy development, negotiations, and implementation	Percentage of industry-specific trade barriers addressed that were removed or prevented[a]	An example of a trade barrier MAS addressed that was removed is an Indian tariff on general-aviation aircraft.
	Percentage of industry-specific trade barrier milestones completed	An example of a trade barrier milestone for removing Indian tariffs on general-aviation aircraft is identifying market-access issues of interest to U.S. industry.
	Percentage of trade agreement milestones completed	An example of a trade agreement milestone is to identify industry-specific opportunities resulting from a particular trade agreement, such as the opportunities for the automotive industry resulting from the Korea-U.S. free trade agreement.
Improve efficiency in the distribution of information and data	Percentage of reduction in per-unit costs of data distribution	Changes in per-minute cost of use of TradeStats Express compared to a baseline.
Conduct studies and analysis to promote U.S. competitiveness	Total annual cost savings resulting from the adoption of MAS recommendations contained in its studies and analysis	MAS estimated that its recommendations on restriction levels for a certain health and safety rule resulted in a savings of $287 million to industry over what had been originally proposed.b

Source: GAO analysis of MAS information and data.

[a] See appendix III.

[b] Generally, MAS uses data on costs before and after its contribution, based on the agency's Regulatory Impact Analysis.

It uses the decrease in costs after MAS's suggested change to measure the value of its impact on the regulation.

Table 6. MAS Objectives and Internal Performance Measures to Enhance U.S. Competitiveness in International Markets

Objective	Performance measure	Description
Enhance public and private partnerships	Total annual exports generated by the Market Development Cooperator Program public/private partnerships	The measure includes, among other things, the estimated exports attributable to the partnership's activities.
	Exports generated annually by export-trading companies	Each export-trading company submits an annual report that includes data on the exports it generated.

Source: GAO analysis of MAS information and data.

MAS Does Not Systematically Obtain Feedback from All Its Clients or Track Its Contributions to Policy Decisions Not Covered by Its Performance Measures

MAS does not systematically obtain feedback on its performance from all its clients on whether or not it is meeting their individual needs. In previous work, we stated that it is important for agency managers to collect performance data that are sufficient to support decision making.[14] A MAS official stated it does not have the resources to systematically survey its clients about their satisfaction with MAS's activities. In addition, MAS officials stated they do not view administering a survey to obtain feedback as feasible since many of their clients are senior policy makers. MAS officials said that if their clients were not satisfied with their work, the clients would not continue to request analysis and information.

Officials stated they do obtain feedback from its ITA clients through periodic meetings with managers in the other ITA business units. Although MAS does not obtain systematic feedback from all its clients, officials from other government agencies told us that while there is a great deal of industry expertise in critical areas within MAS, the level of expertise varies by sector, and some gaps do exist. Officials from the Departments of State and Treasury stated that MAS provided valuable analysis on export financing for airplanes for the Organisation for Economic Cooperation and Development interagency delegation on export financing. In addition, officials from the Department of Transportation stated that MAS's knowledge of trade policy complemented its transportation-policy expertise and its technical understanding of engineering and infrastructure, allowing it to better serve industry. However, officials from State and USTR noted that the level of expertise in MAS is not consistent across all sectors and, consequently, the usefulness and relevance of MAS input can be uneven. In addition, an Import Administration official stated that while it relies on MAS to assist with industry analysis of foreign trade zone applicants, the coverage of some sectors is uneven. MAS officials recognized there is unevenness in the level of expertise across sectors and said they are trying to address the issue as part of its internal-review process and by providing better training for analysts.

Some of MAS's activities fall outside its external and internal measures. For example, MAS officials stated it is difficult to measure some of the economic analysis work that does not contribute directly to breaking down trade barriers or toward developing a trade agreement. Some of this work is used by senior government officials in Commerce and other parts of the Executive Branch. Further, MAS activities that contribute to the performance measures of other parts of ITA are reported not by MAS, but by the unit with primary responsibility for each specific performance measure. MAS's strategic plan includes performance measures related to the goals and objectives that other ITA units have primary responsibility for and on which it does not report. As a result, the performance measures MAS reports on do not include its contributions to the other ITA units' goals. For example, MAS's contributions to antidumping and countervailing duty proceedings are reported by Import Administration.

Responding to our request, MAS officials compiled two large volumes of examples (i.e., analysis documents, information used in trade negotiations, and regulatory cost analyses) illustrating MAS activities over the past few years. MAS officials told us this was not an exercise the office conducts on a regular basis, and that it was very useful. The officials said that it provided a way for staff to link their activities to broad outcomes separately from the overall performance measures. However, they added that the exercise was resource intensive, and they do not have plans to regularly repeat this type of effort.

CONCLUSION

At a time of heightened interest in both expanding U.S. exports and streamlining government operations to lower costs and avoid inefficiencies, the role and effectiveness of MAS within ITA is of particular interest. Created out of an existing Commerce unit in 2004 to more explicitly focus on U.S.

competitiveness, MAS's services are primarily provided to government clients. Unlike other Commerce offices such as Commercial Service and the Import Administration, or independent agencies such as USITC, MAS functions largely as an internal consulting group for government trade policy. MAS's clients report that MAS provides analysis not readily available elsewhere in the government.

Because of the reduced visibility that accompanies this role, it is important for MAS to clearly define its contribution to the Administration's trade and economic policy. We found that MAS continues to refine its mission and develop decision criteria to prioritize its activities, and enhancements are still undergoing departmental review. Given that MAS does not face a market test for its services, a clear set of priorities could help it meet its clients' needs.

MAS faces challenges in measuring its contributions to trade policy, because as a policy support organization, MAS may have difficulty isolating its specific contributions to the trade policy process. MAS has established a detailed process to measure its performance, but the office does not systematically obtain feedback from all its clients or track its contributions to major policy decisions that fall outside its performance measures. This makes it difficult for MAS to accurately determine the extent to which it adds value or to identify opportunities for improvement. In terms of communication, MAS also faces challenges in creating transparency because of the nature of its activities. Because MAS provides industry and trade policy expertise, seldom through products that are publicly attributable to MAS, there is limited visibility of MAS's contributions in key work areas. Its Web site could be a useful communication tool, but as part of ITA, MAS relies on ITA-level support for management of the Web site. Nonetheless, since its support functions are central to its contributions, presenting them more clearly would be useful for congressional and stakeholder oversight of its activities. MAS is in the process of finalizing its mission statement and decision criteria and is taking steps to more clearly prioritize its activities and better align its resources to meet the goals of the Administration's National Export Initiative. Moving this initiative forward within Commerce will be an important step.

RECOMMENDATIONS FOR EXECUTIVE ACTION

To better assure MAS is meeting the needs of its clients, we recommend that the Secretary of Commerce, in concert with MAS management, take the following four actions:

1) To facilitate MAS's efforts to prioritize its activities, establish time frames to finalize the clarification of MAS's mission and decision criteria.
2) To better enable MAS to target its resources, ensure MAS has a way to more systematically monitor how staff time is allocated across various efforts.
3) To improve transparency and ensure that priorities are consistent with those of key stakeholders, explore methods for MAS to more clearly communicate its mission, priorities, and activities to clients, stakeholders, the public, and Congress. These methods could include, among others, working with ITA leadership to develop a strategic plan with associated time frames to improve ITA's web presence and management.
4) In order to ascertain whether MAS is meeting the needs of its government clients involved in the trade policy process, explore ways to more systematically obtain information on the value it is adding. This could include collecting feedback from its clients on its activities more systematically and tracking the outcomes of the analyses it provides for major trade policy decisions.

Agency Comments and Our Evaluation

We requested comments on a draft of this report from Commerce. In its comment letter, Commerce stated that it fully concurred with our findings and recommendations. The letter also said that MAS's current redefinition plan is expected to be completed by October 2011.

Loren Yager
Director,
International Affairs and Trade

Appendix I: Objectives, Scope, and Methodology

The objectives of this report were to examine (1) the Office of Manufacturing and Services's (MAS) goals and activities, and how the types of analysis and expertise it provides compare with those provided by other government entities; (2) how MAS prioritizes its activities and targets its resources; and (3) the extent to which MAS tracks and reports its contributions to increasing U.S. competitiveness and trade.

To assess MAS's activities and how they compare with the types of analysis and expertise provided by other government entities, we reviewed relevant documents, including legislative authority, Department of Commerce's (Commerce) 2004 *Manufacturing in America* report, National Export Initiative documents, and budget and staffing information. We reviewed the International Trade Administration's (ITA) strategic plan and Web site to evaluate how the type of work MAS undertakes compares to that conducted by the other ITA business units. Additionally, we reviewed information on other agencies' Web sites to obtain information on the types of analysis and expertise they offer and initiatives the agencies participated in with MAS. We discussed similarities and differences in the types of work that MAS's government clients conduct with officials with knowledge of the unit. We interviewed Commerce officials from the Office of the Secretary and National Institute of Standards and Technology and representatives from the other ITA business units—Commercial Service, Market Access and Compliance, and Import Administration. We also interviewed officials from other U.S. government agencies, including International Trade Commission, Office of the United States Trade Representative, the Departments of State, Energy, Transportation, Treasury, Environmental Protection Agency, and Office of Management and Budget (OMB). Additionally, we interviewed the Industry Trade Advisory Committees' designated federal officers (DFO), representatives from two industry associations, and the chairman of one of the committees. To gain additional information on MAS's activities, we also attended multiple MAS events, including its Manufacture America conferences in Pittsburgh, Pennsylvania, and Chicago, Illinois; a meeting of the U.S. government's Manufacturing Council; and a congressional event in Washington D.C., for ITA's Market Development Cooperator Program. We also reviewed documents provided by MAS containing examples of how its work has been used by other parts of Commerce and other U.S. government agencies over the last 5 years.

To obtain information on how MAS prioritizes its activities and targets its resources, we reviewed documents related to MAS's proposed decision-making criteria and revised mission and discussed them with MAS officials. Additionally, to assess the availability of information to the public about MAS's role in the U.S. government, we reviewed the Commerce Inspector General's 2007 report on trade coordination efforts[1] and met with ITA's Web

Governance Board. In order to understand any past and current issues related to MAS's priorities, we reviewed OMB's 2006 *Program Assessment Rating Tool* (PART) report and National Export Initiative documents. We also interviewed MAS leadership from the offices of the Assistant Secretary and Planning, Coordination, and Management; officials from its offices of Manufacturing, Services, and Industry Analysis; the administrators of MAS's assignment tasker system; and ITA's Deputy chief financial officer.

To assess the extent to which MAS tracks and reports its contributions to increasing U.S. competitiveness and trade, we reviewed documents related to its performance measurement system, including MAS's industry assessments, business plans, performance targets, and Commerce's fiscal year 2010 *Performance and Accountability Report*. We also reviewed past GAO work on the principles of effective performance measurement. In order to ensure consideration of any past issues concerning MAS's performance measurement, we reviewed OMB's 2006 PART report. We also discussed MAS's performance measurement with knowledgeable officials from MAS, ITA, and OMB.

In order to present information on manufacturing's share of U.S. gross domestic product (GDP), employment, and exports, we relied on data from Commerce. Information on GDP and employment was obtained from the Bureau of Economic Analysis (BEA). We used calculations of hours worked to report shares of employment. To report the numbers of employees in the U.S. economy, we computed full-time equivalents, assuming 2,080 hours worked in each calendar year. Information on manufacturing exports was obtained from MAS's TradeStats Express. To compute the U.S. share of world-wide production and exports in manufacturing data, we relied on the World Bank's Data Bank. To assess the reliability of the World Bank and Commerce data, we interviewed knowledgeable agency officials. To the extent possible, we compared values against alternative sources. We found that the data were sufficiently reliable for the purposes of presenting historic trends on manufacturing production, exports, and employment.

We conducted this performance audit from August 2010 to June 2011 in accordance with generally accepted government auditing standards. Those standards require that we plan and perform the audit to obtain sufficient, appropriate evidence to provide a reasonable basis for our findings and conclusions based on our audit objectives. We believe that the evidence obtained provides a reasonable basis for our findings and conclusions based on our audit objectives.

APPENDIX II: MANUFACTURING'S ROLE IN THE U.S. ECONOMY AND EXPORTS

Manufacturing's Share of U.S. GDP Has Declined, but the United States Remains a Leading Manufacturer

Over the past two decades, manufacturing's share of the output of the U.S. economy has fallen.[1] In 1989, manufacturing's output expressed in constant 2009 dollars to correct for inflation was approximately $1.5 trillion, while total output was approximately $8.6 trillion.[2] By 2009, while manufacturing output had grown to $1.6 trillion, total output had grown to more than $14 trillion. Using data from BEA, figure 3 shows manufacturing's share of total U.S. output, measured by GDP. As the figure shows, from 1989 to 2009, U.S. manufacturing as a share of GDP declined from a peak of about 17 percent to about 11 percent, and has generally displayed a steady decline.[3]

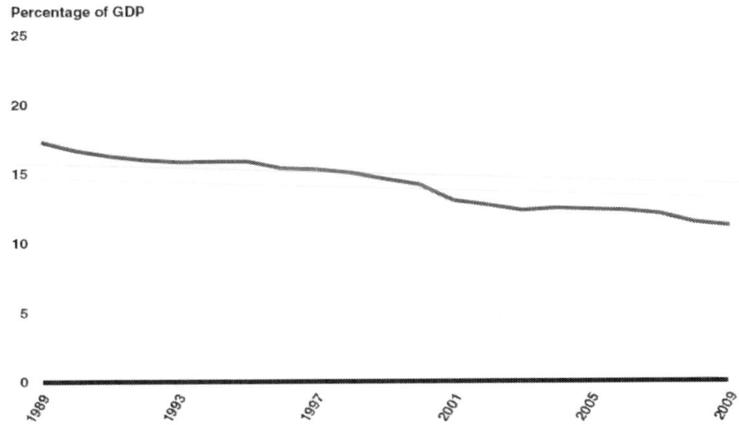

Source: GAO analysis of BEA data.

Note: Figure shows the ratio of the value added in Manufacturing over nominal GDP. Correcting for inflation would not change the share if the same inflation factor were applied to the numerator (manufacturing GDP) and the denominator (total GDP). However, analysts sometimes adjust manufacturing output by a different inflation factor that reflects slower rises in the prices of manufacturing goods and productivity gains. Using that inflation factor, manufacturing's share of real output would have remained roughly constant over this period.

Figure 3. U.S. Share of Nominal GDP Represented by Manufacturing, 1989 – 2009.

While the share of U.S. economic output devoted to manufacturing has fallen, the United States remains a leading producer of manufactured goods. Table 7 uses data from the World Bank to compare selected countries' proportions of world production in manufacturing in 2000 and 2008. These four countries collectively represent more than 50 percent of world manufacturing. As the table shows, the U.S. share of world production in manufacturing fell from about 26 percent to about 18 percent. During the same period, China's share increased from about 7 percent to about 15 percent.

Table 7. Value of Manufacturing GDP and Percentage of World Manufacturing GDP for Selected Countries, 2000 and 2008 (Dollars in billions)

	2000		2008	
Country	Value of manufacturing output	Percentage of world manufacturing output	Value of manufacturing output	Percentage of world manufacturing output
United States	$1,815.4	26%	$1,768.0	18%
Germany	485.3	6.8	745.5	7.6
Japan	1,278.7	18	979.3	10
China	476.0	6.7	1,502.2	15
World	$7,089.8	100%	$ 9,840.0	100%

Source: GAO analysis of World Bank data.

Note: Dollar values have been adjusted using the U.S. GDP price deflator, with 2009 as the base year.

Manufacturing Exports Remain a Chief Component of Total U.S. Exports

Over the past 20 years, the percentage of U.S. exports represented by manufacturing has remained largely unchanged. In 1989, correcting for inflation, the United States exported approximately $800 billion of goods and services, of which about $450 billion were manufactured goods, according to BEA and MAS data.[4]

In 2009, the United States exported approximately $1.6 trillion, of which about $917 billion were manufactured goods. Figure 4 shows the share of total exports that were in manufacturing.

Over this time manufacturing's share of exported goods and services has stayed between 58 and 65 percent.

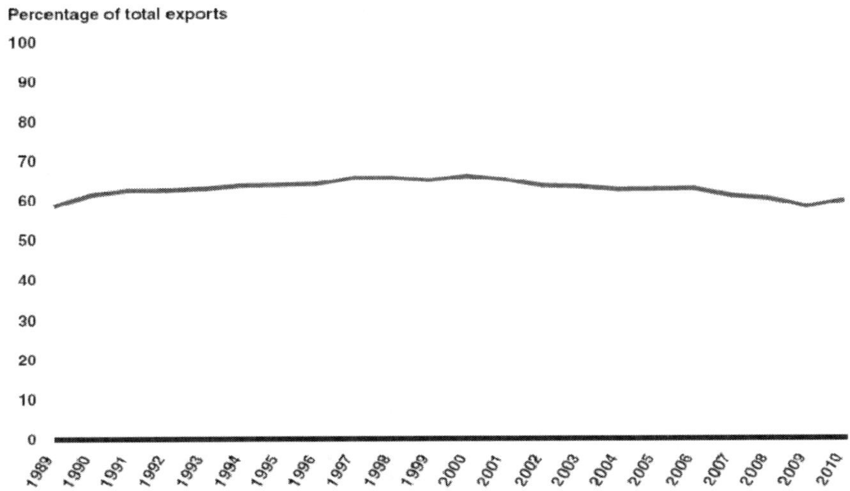

Source: GAO analysis of MAS and BEA data.

Note: Figure shows manufacturing's share of total U.S. exports, calculated using nominal data for export values. If we had presented the share of exports correcting for inflation, the share would be identical because the same inflation factor would apply to the numerator (manufacturing exports) and the denominator (total exports).

Figure 4. U.S. Manufacturing Exports as a Percentage of Total U.S. Exports.

The U.S. share of world-wide exported manufactured goods has remained consistent over the past decade. Table 8 shows manufacturing exports for selected countries in 2000 and in 2009 using World Bank data.[5] In 2000, the United States represented about 13 percent of manufactured merchandise exported, which declined to about 8 percent in 2009. During this period, China's contribution to world manufacturing rose from about 5 percent in 2000 to about 13 percent in 2009.

Over the Past Two Decades, Manufacturing's Share of Employment Has Fallen

From 1989 to 2009, manufacturing's share of employment has fallen. In 1989, measured by hours worked by full- and part-time employees, there were approximately 18 million U.S. workers in manufacturing and approximately 94 million in the entire U.S. economy.[6] In 2009, the number employed in manufacturing had fallen to approximately 11 million, while the number employed in the United States had grown to about 107 million. Figure 5 shows the share of hours worked in manufacturing, which fell from almost 20 percent in 1989 to about 10 percent in 2009.[7]

Table 8. Value of Manufacturing Exports and Percentage of World Manufacturing Exports for Selected Countries, 2000 and 2009 (Dollars in billions)

	2000		2009	
Country	Value of manufacturing exports	Percentage of world manufacturing exports	Value of manufacturing exports	Percentage of world manufacturing exports
United States	$800.0	13%	$705.3	8.1%
Germany	570.9	9.6	918.5	11
Japan	556.3	9.3	511.2	5.9
China	271.8	4.6	1,124.3	13
World	$4,817.5	100%	$ 8,731.6	100%

Source: GAO analysis of World Bank Data.

Note: Dollar values have been adjusted using the U.S. GDP price deflator, with 2009 as the base year.

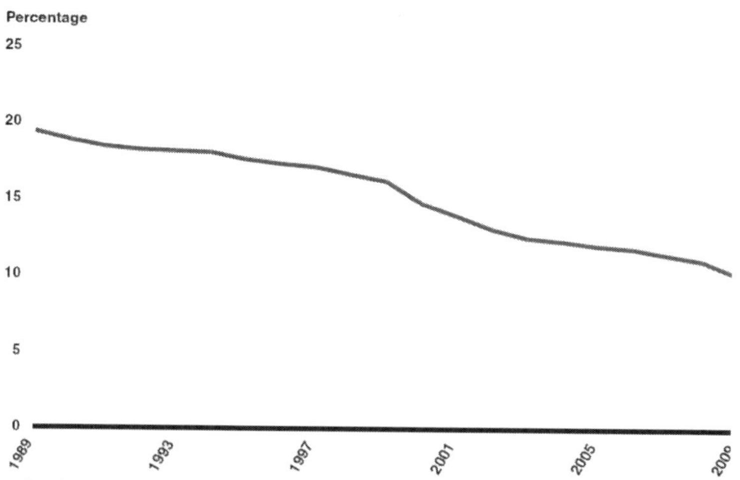

Source: GAO analysis of BEA data.

Note: Figure shows manufacturing's share of total hours worked by full- and part-time employees.

Figure 5. Manufacturing's Share of U.S. Employment.

Over the past 10 years, the drop in manufacturing employment has been matched by an increase in the share of employment in services. Table 9 shows employment in the United States by sector for 2000 and 2009. As the table shows, over this period manufacturing's share has fallen by about 5 percentage points, while the service industry's share has increased by about 4 percentage points.

Table 9. Number and Percentage of U.S. Workers Employed in Various Sectors, 2000 and 2009

(Numbers in millions)				
	2000		2009	
Sector	Employment	Percentage of U.S. employment	Employment	Percentage of U.S. employment
Manufacturing	16.3	15%	10.9	10%
Services	70.1	63.2	72.1	67.2
Other goods	8.4	8	7.6	7
Government	16.5	15	17.6	16.4
Total	110.9	100%	107.2	100%

Source: GAO analysis of BEA data.

Note: Other goods-producing industries are agriculture, forestry, fishing, and hunting; mining; and construction. Services-producing industries consist of utilities; wholesale trade; retail trade; transportation and warehousing; information; finance and insurance; real estate and rental and leasing; professional, scientific, and technical services; management of companies and enterprises; administrative and waste management services; educational services; health care and social assistance; arts, entertainment, and recreation; accommodation and food services; and other services, except government.

APPENDIX III: INDUSTRY-SPECIFIC TRADE BARRIERS

Manufacturing	Services
Access to China's electronics recycling market	China insurance branching and licensing
Anticounterfeit medical products	Japan postal insurance (new)
Asia-Pacific Economic Cooperation (APEC) automotive customs transactions cost reduction	Universal postal union postal financial services
APEC cross-border privacy	U.S.-Japan air services agreement
APEC life sciences innovation forum	
Automotive standards harmonization	

(Continued)

Manufacturing	Services
Canada and Mexico CITEL Mutual Recognition Agreement implementation	
Cement Asia-Pacific	
China automotive trade barriers	
China general aviation tariffs	
China illegal logging	
China Joint Commission on Commerce and Trade health	
China large motorcycle restrictions	
China raw materials	
Environmental trade liberalization World Trade Organization	
European Union (EU) access for U.S. wine (Phase II)	
India general aviation tariffs	
India motorcycle tariffs	
Japan boats	
Japan market access for information technology (IT) and E-Commerce	
Japan mobile competition	
Korea automotive trade barriers	
Medical device global harmonization	
Nuclear industry-specific trade barriers	
Privacy framework standard International Organization for Standardization 29100	
Russia import duty on large civil aircraft	
Trans-Pacific Partnership—telecom services and E-Commerce	
U.S. cybersecurity policy	
U.S. Internet policy on privacy and innovation	
U.S.-EU safe harbor	
U.S.-EU World Trade Organization IT agreement tariff dispute	
Vietnam telecom and e-Commerce	
World wine group (Phase II)	

Source: MAS, Fiscal Year 2010 List of Industry-Specific Trade Barriers (Washington, D.C., 2010).

APPENDIX IV: MAS CREATED A NEW SECTOR STRATEGY DOCUMENT IN RESPONSE TO THE NATIONAL EXPORT INITIATIVE

Recently, in response to the National Export Initiative, MAS began to prepare "global sector strategies," which, according to MAS officials, identify and provide information on potential export markets for high-priority sectors; identify obstacles and risks associated with the sector; and provide policy recommendations for decision makers. Agency officials stated that the global sector strategies are based on a common template and will be MAS's signature product, with portions available to the public.

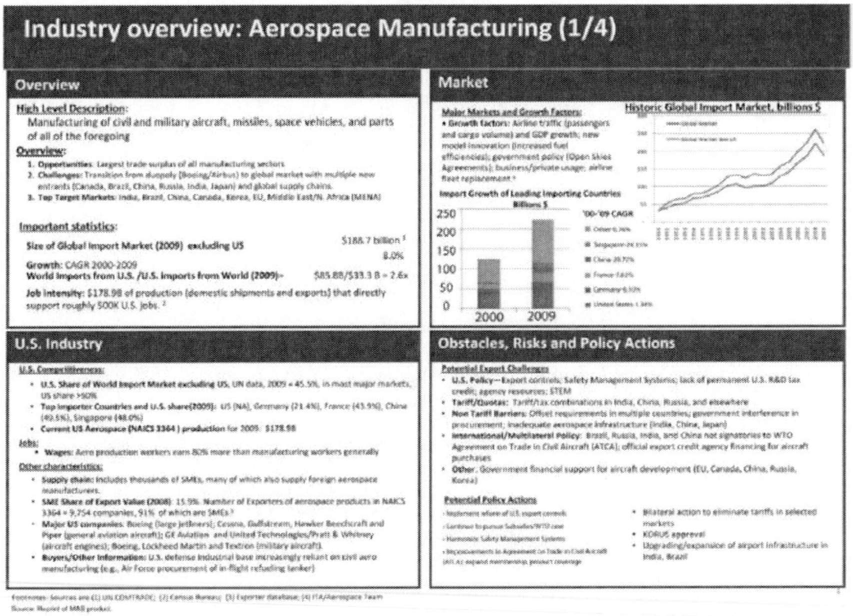

Figure 6. MAS Sector Strategy Example—Excerpt from Aerospace Sector Strategy.

Currently, one strategy, for aerospace, has been completed. An excerpt is reproduced in figure 6. The other 14 strategies that are listed below are in various stages of completion:

- Automotive;
- Basic chemicals;
- Civil nuclear;
- Travel and tourism;
- Professional services;
- Digital content;
- Medical technologies;
- Building products;
- Semiconductor industry;
- Franchising/distribution;

- Renewable energy;
- Supply chain and logistics;
- Architecture, engineering, and construction; and
- Insurance and asset management.

End Notes

[1] Executive Order 13534, Mar. 16, 2010.

[2] ITA has six long-term performance goals to which its business units contribute: (1) advance U.S. international and commercial strategic interests; (2) enhance U.S. competitiveness in domestic and international markets; (3) broaden and deepen the U.S. exporter base; (4) identify and resolve unfair trade practices; (5) foster excellent relationships with customers and stakeholders; and (6) achieve organizational and management excellence.

[3] House Report No. 108-221, 108th Cong., 1st Sess. (2003), pp. 68-70.

[4] Consolidated Appropriations Act, 2004 (Pub. L. No. 108-199, Div. B, Title II, 118 Stat 3 at 65); ITA Organization and Function Order 41-1, Aug. 18, 2005.

[5] Executive Order 13534.

[6] MAS's primary goal aligns with ITA's second long-term goal—enhancing U.S. competitiveness in domestic and international markets.

[7] MAS contributes to some ITA trade promotion efforts by, for example, managing the programmatic aspects of ITA's Market Development Cooperator Program (MDCP), which awards funds annually to industry associations and nonprofit organizations, rather than individual companies. Additionally, in response to the expectation expressed in a House of Representatives report that MAS would redirect industry experts to the field to focus on the needs of local industries, MAS initiated a pilot Field Assignment program in June 2004. MAS sent multiple industry specialists to the field temporarily to cover numerous industries. The program ended in March 2005, and, according to agency officials, was not continued due to competing priorities and resource constraints.

[8] GAO, *Standards for Internal Control in the Federal Government,* GAO/AIMD-00-21.3.1 (Washington, D.C.: November, 1999).

[9] GAO/AIMD-00-21.3.1.

[10] Commerce Office of Inspector General, *International Trade Administration: Commerce Can Further Assist U.S. Exporters by Enhancing Its Trade Coordination Efforts,* Final Inspection Report No. IPE-18322 (Washington, D.C., March 2007).

[11] In 2006, OMB found that MAS's annual strategic planning process links its budget process to goals, objectives, and results.

[12] MAS has both Government Performance and Results Act measures, which it reports externally, and performance measures that it uses internally but does not report externally. These objectives and performance measures were initially laid out in MAS's strategic plan for fiscal years 2007-2012, created in preparation for the OMB Program Assessment Rating Tool (PART) review in 2006.

[13] Although, as previously noted, MAS has five performance measures related to the Government Performance and Accountability Act, Commerce's *Performance and Accountability* report does not always include results for all five measures. While the reports for fiscal years 2006 and 2007 included the measure on reduction in costs of data distribution, it has not been

included in subsequent *Performance and Accountability* reports, although MAS officials stated they continue to track this measure and report on it to OMB.

[14] GAO, *Executive Guide: Effectively Implementing the Government Performance and Results Act,* GAO/GGD-96-118 (Washington, D.C.: June 1996).

End Note for Appendix I

[1] Commerce, Office of Inspector General, *International Trade Administration: Commerce Can Further Assist U.S. Exporters by Enhancing Its Trade Coordination Efforts,* Final Inspection Report No. IPE-18322 (Washington, D.C., March 2007).

End Notes for Appendix II

[1] According to the North American Industrial Classification System, the manufacturing sector comprises establishments engaged in the mechanical, physical, or chemical transformation of materials, substances, or components into new products. Establishments in the manufacturing sector are generally plants, factories, or mills that characteristically use power-driven machines and materials-handling equipment.

[2] The trend of the past 20 years is consistent with a larger trend. In 1969, the total value of manufacturing output was about $ 1.1 trillion, or about 24 percent of the entire economy's $4.7 trillion. Manufacturing output has been adjusted to reflect 2009 prices, using BEA's GDP price deflator.

[3] Declining shares of manufacturing in GDP are not unique to the United States. According to data from the World Bank, manufacturing's share of world GDP was about 20 percent in 1993, and about 16 percent in 2009. For calculating manufacturing GDP, the World Bank uses industries belonging to International System of Industrial Classification codes 15-37.

[4] Figures have been adjusted for inflation using BEA's GDP price deflator with 2009 as the base year.

[5] The World Bank defines manufacturing using the Standard International Trade Classification system. According to MAS officials, the World Bank data is more limited, as it excludes processed food and petroleum products.

[6] Figures are full-time equivalents that assume a work year of 2,080 hours.

[7] In 1969, manufacturing's share of U.S. employment was approximately 28 percent.

INDEX

F

G

H

I

J

K

L

M

N

Q

R

S

T